International bidding case study

International Construction Management Series No. 2

International bidding case study

Andrew Baldwin
Ronald McCaffer
Sherif Oteifa

International Labour Office   Geneva

Baldwin, A., McCaffer, R. and Oteifa, S.
International bidding case study
Geneva, International Labour Office, 1995 (International Construction Management Series No. 2)
/Textbook/, /Model/, /Project management/, /Cost accounting/, /Construction/, /Construction industry/. 08.10.1
ISBN 92-2-108268-7
ISSN 1020-0142

ILO Cataloguing in Publication Data

Printed in the United Kingdom   EPL

D
692.5
BAL

PREFACE

The International Labour Office, through its Enterprise and Cooperative Development Department, has a continuing interest in the development of construction enterprises throughout the world. Specifically, this interest stems from the dual role of construction not only as a significant source of direct employment but as a sector which contributes—through its wide range of operations and projects—to the growth and development of virtually all other economic sectors.

Many construction enterprises have chosen to limit their operations to the home market. Others appreciate the scope for diversifying into international projects, but are wary of the risks involved in venturing into an alien environment. Others still have suffered severe losses as a result of international ventures, and have resolved to avoid them in future. The caution regarding international contracting is well-founded, but the potential attractions remain. The world construction market is very large, and the projected demand for improved infrastructure and shelter suggests that it will continue to offer attractive opportunities to the discriminating bidder. Many national contractors could offer a competitive service in selected areas, provided they offer realistic bids based on a proper appreciation of resources and risks.

Books in the International Construction Management (ICM) series have been written for engineers and members of other construction professions who are working in such enterprises, and wish to fit themselves for a career in international project management. They are also suitable for undergraduate and postgraduate engineering students. Such readers are likely to have a technical background in civil engineering or a related discipline, but may have only a limited understanding of accounting, finance, strategic management, marketing and commercial topics. Whilst there are numerous books that tackle the latter topics in a national context, the ICM series seeks to provide an integrated introduction to international construction management as a basis for making informed investment and operating decisions so as to minimize risk and improve the profitability of the construction enterprise.

The ICM series is ideal both for individual study and group study, since relatively complex topics are carefully explained and the active participation of the reader is ensured by the extensive use of case-studies, worked examples and exercises. Although United Kingdom practice is used as a frame of reference, the books encourage the reader to compare alternative approaches and adapt the application of techniques to suit local regulatory, contractual and tax regimes.

The text of *International bidding case study* is divided into four chapters, and takes the reader through the various stages of bidding for a major international project. The first chapter sets out the contract documents in the form that they would be received by a prospective bidder. The second chapter contains a commentary by the bidder on the documents, setting out the considerations that would affect the decision as to whether to bid and the way in which the site work would be organized. The third chapter covers the estimation of construction costs, leading to the decision on the tender price and the submission of the tender in Chapter 4.

Overall the book presents a unique insight into the way in which an experienced international contractor approaches a bidding opportunity, in this case a new harbour in the (hypothetical) Republic of Federalstan together with associated marine works. The finance for this project has been arranged through the (equally hypothetical) Euroasian Development Bank, while the contractor who is preparing the tender is based in a hypothetical country known as the Unified Republic which is a considerable distance from the location of the new project in Federalstan.

The preparation of a tender for an international capital project inevitably involves the consideration of several different currencies. For the purposes of the case study it is assumed that the main currency of the tender is dollars, the currency of the Unified Republic (UR). The national currency of the Republic of Federalstan (RF) is the Nu. Thirty (30) Federalstani Nu are equivalent to one UR dollar.

The authors emphasize that, although the case study provides a realistic insight into the international bidding process, all details such as location, project design, finance, currencies and unit costs are wholly hypothetical. Any relationship to existing or proposed projects and the parties involved in these projects is purely coincidental.

This book has been prepared by Andrew Baldwin, Professor Ronald McCaffer and Dr. Sherif Oteifa of the Loughborough University of Technology in the United Kingdom and edited by Derek Miles, Director of Overseas Activities in the Department of Civil Engineering of Loughborough University of Technology, United Kingdom, and previously Coordinator of the ILO Construction Management Programme.

ACKNOWLEDGEMENTS

This book in the International Construction Management (ICM) series was designed to assist the China International Contractors' Association (CHINCA) to meet the needs of its member corporations to build the broad management expertise needed to expand their operations on the international market. It was prepared within a technical cooperation project carried out by the ILO as executing agency under the United Nations Development Programme (UNDP). In view of the dearth of textbooks and training material that would be suitable for the needs of senior and middle-level management staff of these enterprises, it was decided to commission a series of texts that would enable them to improve their understanding of the range of specialist management skills that are required to compete effectively in the international construction market.

We are grateful to the staff of CHINCA and its member corporations for their advice and assistance during the field testing of the material upon which this book is based, and the agreement of both CHINCA and UNDP that it should be made accessible to a wider audience.

CONTENTS

Figures

THE NEW HARBOUR PROJECT CONTRACT DOCUMENTS 1

The documents in this chapter will be the subject of both a preliminary appraisal and a detailed review.

A suitable check-list should be produced by the estimator and used to check these legal documents carefully to assess in full their implications.

1.1 Instructions to Tenderers

The following instructions are provided to each of the tenderers for the project. They are divided into the following sections:

(a) Source of funds; (b) Eligibility requirements; (c) Scope of work; (d) Pre-tender site visit; (e) Tender Documents; (f) Supplementary information to Tenderers; (g) Clarification of Tender Documents; (h) Preparation of the Tender; (i) Submission of the Tender; (j) Tender Prices; (k) The currencies for the Tender and for payment; (l) Alternative designs; (m) The Bid Bond; (n) The validity of the Tender; (o) Communication; (p) Submission of Tenders; (q) Opening of Tenders; (r) Contacting the Employer; (s) Contract award criteria; (t) Notification of Intent of Award; (u) Signing of the Contracts; (v) Performance Bond; (w) Disbursement.

(a) Source of funds

The Republic of Federalstan has received a loan from the Ordinary Funds Resources of the Euroasian Development Bank (EDB) in various currencies towards the cost of the New Harbour Project. It is intended that the proceeds of this loan (Loan No. 1000-91) will be applied to eligible payments under the Contract for the construction of this Cargo Jetty and Breakwater. The terms and conditions of the Contract and payments will be subject to the terms and conditions of the loan agreement including the *Guidelines for Procurement under EDB Loans*.

Except as the Bank specifically agrees, no party other than the Republic of Federalstan shall derive any rights from the loan agreement or have any claim to the loan proceeds.

(b) Eligibility requirements

The foreign exchange costs and part of the local cost of the Contract are intended to be financed from a loan from the Special Funds Resources of the Euroasian Development Bank. To permit such financing, the Contract must meet the following eligibility requirements:

(i) the Contractor and any subcontractor must be a national of an eligible source country;

(ii) the services supplied under the Contract must be wholly or substantially supplied from an eligible source country or countries; and

(iii) any material and equipment supplied by the Contractor under the Contract must be procured in an eligible source country or countries.

For the purposes of these requirements, the term *eligible source country* means any member country of the Euroasian Development Bank as listed on the list of Eligible Countries held at any Embassy of the Republic of Federalstan.

(c) Scope of work

The Federalstan Harbour Authority (FHA), hereinafter referred to as the Employer, invites pre-qualified contractors (hereinafter referred to as the Tenderers) to submit Tenders for the Contract, which covers the construction of a new unloading jetty and breakwater.

A brief description of the scope of the work for this Contract is made below:

(i) site clearance;

(ii) new unloading jetty (105 m long and 20 m wide);

(iii) breakwater (caisson type, 600 m long);

(iv) dredging (3,000,000 m^3);

(v) reclamation (100,000 m^3);

(vi) slope protection (100 m).

The works shall be completed, including mobilization, preparatory work and demobilization, in 18 months or less from the commencement of the works.

(d) Pre-tender site visit

The Tenderers shall inspect and examine the site of the proposed works and its surroundings and obtain for themselves at their own responsibility all information that may be necessary for the purpose of making a tender and entering into the Contract.

All costs and charges involved in their visit to the site shall be borne by themselves.

The Tenderers or their representatives shall so long as they notify the employer in writing be granted permission on application to visit and inspect the site on any days before the closing date of the Tender.

(e) Tender Documents

The Tender Documents comprise the following:
- Instructions to Tenderers;
- Form of Tender;
- Form of Bid Bond;
- Form of Agreement;
- Conditions of Contract;
- Technical Specifications;
- Bill of Quantities;
- Drawings.

(f) Supplementary information to Tenderers

Supplementary information and data furnished by the Employer will be made available on request to Tenderers for reference during the preparation of the Tender. The Employer does not guarantee and is not responsible for the accuracy of the information and data. The Tenderers shall accept any responsibility for interpretation, evaluation and conclusion on the information including geological and weather conditions.

(g) Clarification of Tender Documents

Tenderers may request the Employer in writing, copied to the Engineer up to 14 days before the Tender closing date, for clarification of any aspect of the Tender Documents. Any clarification by the Employer or Engineer shall be delivered in writing to every Tenderer. Where such clarification requires amendment in the Tender Documents, the Employer or Engineer shall issue an Addendum to the Documents and this shall be part of the Tender Documents, binding on Tenderers.

(h) Preparation of the Tender

The Tenderer shall sign and date each of the Documents in the space provided for this purpose. The Tenderer and all correspondence and documents shall be written in the English language, or accompanied by a Federalstani language translation. Where any discrepancy exists between these languages, the Federalstani language shall govern for the purpose of interpretations of the Tender.

Erasure or defacement of the Schedules will not be permitted. Any revision of prices and other conditions in the Documents shall be made in figures and words, and shall be accompanied by a signature.

(i) Submission of the Tender

The Tender submission shall comprise the following:

(i) Form of Tender;

(ii) Bill of Quantities; and

(iii) Bid Bond, duly completed by the Tenderer's Guarantor.

The completed Tender shall have no interlineations or erasures, except those required to correct errors made by the Tenderer in which case those shall be initialled by the person signing the Tender.

(j) Tender Prices

(i) The Tenderer shall fill in rates and amounts for all items of works described in the Bill of Quantities, whether quantities are stated or not. Items not specified in the Bill of Quantities but considered by the Tenderer to be essential and necessary for the completion of the works shall be deemed to be covered by the other unit or lump-sum Tender Prices in the Bill of Quantities and will not be paid for separately.

(ii) The terminology of unit price and schedule rate will be used in the same meaning.

The Tenderer must calculate his unit prices for the different items of works specified in the Bill of Quantities, and write these prices in the *Schedule rate* column.

The Tenderer shall multiply these schedule rates by the quantities indicated in the Bill of Quantities and write the results in the *Amount* column.

It should be noted that the above amounts are solely required for the purpose of evaluating and comparing various tender proposals and shall not be deemed to be actual sums to be paid to the Contractor.

The actual sums to be paid to the Contractor shall be determined by measuring the works actually done in accordance with the terms of the Contract.

(iii) There are some items which require the Tenderer to insert the Tender Prices on a lump-sum basis. This means that the Tenderer must calculate the Tender Prices at a fixed lump-sum price including all costs which are deemed essential to execute an item of works or services specified in the Contract Documents. The items priced on such a lump-sum basis shall be paid at the same price, irrespective of the actual increase or decrease in quantity of the works.

(iv) The Tender Prices are deemed to include all direct and indirect expenses, whether local or foreign components, and all incidental and contingent costs, profits, overheads, allowances, taxes, duties, risks, insurance and obligation of every kind which must be borne by the Contractor since they are necessary to complete the whole works in accordance with the Contract.

(v) It shall be noted that any unpriced item in the Bill of Quantities will not relieve the Tenderer of his obligation to complete the works under this item once contracted and that no payment will be made against any such unpriced item.

(vi) Should any discrepancy be found in the proposed Tender Prices, the following priority of handling will be applied:

- in case of discrepancies between schedule rate and extended amount, the schedule rate shall prevail;

- in case of discrepancies between the actual sum of the total Tender Prices for individual items and the Overall Total Tender Price, the actual sum for individual items shall prevail.

(k) The currencies for the Tender and for payment

(i) The unit rates and prices shall be quoted by the Tenderer entirely in Federalstani Nu. A Tenderer expecting to incur expenditure in other foreign currencies for inputs to the Works supplied from outside the Employer's country shall indicate the percentage of the Tender Price needed for the payment of such foreign currency requirements, either:

- entirely in the currency of the Tenderer's home country or at the Tenderer's option;

- in the currency of another stated country;

always provided that a Tenderer expecting to incur expenditures in a currency or currencies other than those stated above for a portion of the foreign currency requirements, and wishing to be paid accordingly, shall so indicate the percentage portion in the Tender. The amounts in various currencies, calculated on the basis of the percentages indicated in the Tender and by the use of the exchange rates indicated in the sub-clause (ii) hereinafter, shall be used for the purposes of payment.

(ii) The exchange rates to be used by the Tenderer for currency conversion shall be the selling rates for similar transactions published by the State Bank of Federalstan, prevailing on the date thirty (30) days prior to the latest date for the submission of Tenders. If exchange rates are not so published for certain currencies, the Tenderer shall state the exchange rates used and the source for the purpose of payments, the exchange rates used in the Tender preparation shall apply for the duration of the Contract.

(l) Alternative designs

Tenderers may propose alternative designs for all or part of the works. Any such alternative designs must be fully supported with a detailed priced Bill of Quantities, detailed drawings, any additional specifications, complete design calculations and proposed method statement of the construction. Failure to either submit a Tender in accordance with the exhibited design, or to provide full supporting documentation in the event that an alternative design is proposed by the Tenderers will automatically disqualify the Tender. Alternative Tenders must be accompanied by their own separate Bid Bond and each alternative Tender will be evaluated on the same basis as the Main Tenders.

(m) The Bid Bond

The Tenderer shall provide a Bid Bond as security for the Tender equivalent to five per cent (5 per cent) of the Total Tender Sum. The Bid Bond shall be denominated in Federalstani Nu or in a freely convertible currency and shall be in one of the following forms:

   (i)   bank guarantees issued by a reputable bank in the form provided in the Tender Documents or another form acceptable to the Employer and valid for thirty (30) days beyond the validity of the Tender; or

   (ii) cashier's cheque, certified cheque or cash.

Any Tender not secured in accordance with clause (m)(i) and (ii) above will be rejected by the Employer.

A successful Tenderer's Bid security will be returned as promptly as possible on award of Contract, but in any event not later than thirty (30) days after the expiration of the period of Tender validity prescribed by the Employer, pursuant to clause (n) hereof.

The successful Tenderer's Bid security will be returned upon the Tenderer's executing the Contract, pursuant to clause (u), and furnishing the Performance Bond, pursuant to clause (v) hereinafter.

The Bid Bond may be forfeited:

– if the Tenderer withdraws his Tender during the period of Tender validity; or

– in the case of a successful Tenderer, if the Tenderer fails:

   (i)   to sign the Contract in accordance with clause (u);

   (ii) to furnish the Performance Bond in accordance with clause (v).

(n) The validity of the Tender

The Tender shall remain valid and open for acceptance by the Employer for one hundred and twenty (120) calendar days measured from the date appointed for submission of Tender. The Employer may solicit the Tenderer's consent to an extension of the period of Tender validity. The request and the response shall be made in writing. If the Tenderer agrees to the extension request, the validity of the Tender bond shall also be extended. A Tenderer may refuse the request without forfeiting his Bid Bond. A Tenderer granting the request will not be required or allowed to modify his Tender.

(o) Communication

All correspondence by the Tenderer during the Tendering period shall be made by telex addressed to the Employer and copied to the Engineer, as follows:

Employer: To the Federalstan Harbour Authority
Telex No.:   876 FED
Telephone No.:   + 967 208 0020
Attention: Managing Director

Engineer: To Newtown Engineering Consultants
Telex No.:   0001 LEC
Telephone No.:   100000 Newtown
Attention: Managing Partner
Civil Engineering Road, Newtown, UR.

(p) Submission of Tenders

Tenders shall be delivered to the Employer at the address and by the time given in the Instructions to Tender.

Any Tender received by the Employer after the deadline for submission of Tenders will be rejected and returned without opening to the Tenderer.

(q) Opening of Tenders

The Employer will open the Tenders, in the presence of the Tenderer's representatives who choose to attend at the time specified in the *Instructions to Tender* and in his or her office.

(r) Contacting the Employer

No Tenderer shall contact the Employer on any matter with regard to his Tender, from the time of Tender opening to the time the Contract is awarded.

(s) Contract award criteria

The Employer will award the Contract to the successful Tenderer whose Tender has been determined by the Employer to be substantially responsive and has been selected by the Employer as the lowest evaluated Tender.

(t) Notification of Intent of Award

The Employer will notify the successful Tenderer in writing that the Tender has been accepted and the basis on which the Tender has been accepted. The notification of *Intent of Award* will constitute the formation of a Contract, until the execution of the formal Contract. Upon the successful Tenderer providing a Performance Bond, the Employer will promptly notify unsuccessful Tenderers that their Tenders have been unsuccessful and their Bid Bond will be returned.

(u) Signing of the Contracts

At the time of notification of *Intent of Award*, the Employer will send the successful Tenderer the Contract Form provided in the Tender Documents, incorporating all agreements between the parties.

Within fourteen (14) days of receipt of such Contract Form, the successful Tenderer shall sign and date the Contract and return it to the Employer.

(v) Performance Bond

Within thirty (30) days of receipt of the *Notification of Intent of Award* from the Employer, the successful Tenderer shall furnish the Performance Bond, in accordance with the Conditions of Contract.

Failure of the successful Tenderer to lodge the required Performance Bond shall constitute sufficient grounds for the annulment of the award and forfeiture of the Bid Bond, in which event the Employer may award the Contract to another Tenderer or call for new Tenders.

(w) Disbursement

If requested by the Employer during the evaluation of Tenders or prior to the award of Contract, Tenderers shall provide an estimate of the quarterly disbursement (excluding Provisional Sums) to be made by the Employer during the period of the Contract.

The estimate of disbursement shall not form part of the Contract.

1.2 Conditions of Contract

The Conditions of Contract for a project will normally be presented as a set of General Conditions of Contract supplemented by a list of conditions particular to the project in question. The General Conditions of Contract may be those prepared by the client or an independent party such as the *Conditions of Contract (International) for Works of Civil Engineering*

Construction, 4th edition, 1987, produced by Fédération Internationale des Ingénieurs-Conseils (FIDC). For the purposes of this case study the following key clauses have been identified for the project.

Definitions and interpretation

Employer means the Federalstan Harbour Authority.

Engineer means the Newtown Engineering Consultants.

Contract Price means the total sum stated in the Agreement, subject to such additions thereto or deductions therefrom as may be made under the provisions hereinafter contained.

Contract Documents

These Contract Documents are drawn up in the English language. The English language shall be the ruling language for all matters and for all communications between the parties.

The Contract is to be construed in accordance with the Law of Federalstan.

Further Drawings and instructions

The Construction Drawings issued by the Engineer do not necessarily show all details of the Works, and the Contractor shall prepare detailed Working Drawings.

The Contractor shall submit, at least twelve (12) weeks before any part of the Works is commenced, for the Engineer's approval three (3) copies of all Working Drawings. Any part of Works shall not be commenced before the relevant Working Drawings are approved by the Engineer.

Performance Bond

The Contractor shall provide within thirty (30) days after the receipt of Employer's *Notification of Intent of Award*, a Performance Bond to the amount of ten per cent (10 per cent) of the Contract Price. Such Bond shall guarantee the performance, completion and maintenance of the Works and shall be issued by a Bank or Bonding Company, as the case may be, acceptable to the Employer. The cost of the Bond shall be borne by the Contractor.

The Bond shall be valid until thirty (30) days after issue of the Maintenance Certificate by the Engineer.

Third party insurance

The minimum amount of the Contractor's third party insurance shall be 60 million Nu per occurrence.

The Employer and the Engineer shall be named as additional insured on the insurance.

Restrictions on working hours

None of the Works, with the exception of the dredging work or other marine works which specifically require the use of floating plant and convenient tide, shall be carried on during the night or on locally recognized days of rest without the consent of the Engineer, except when work is unavoidable or absolutely necessary for the saving of life, property or the safety of the Works.

Liquidated damages for delay

The amount of liquidated damages payable in respect of delayed completion of the Works shall be as follows, each percentage relating to the Contract Price.

Half a per cent (0.5 per cent) for each week of delay or any part thereof, provided that the total amount of liquidated damages shall not exceed ten per cent (10 per cent) of the Contract Price.

Period of maintenance

The period of maintenance shall be twelve (12) calender months from the date of completion of the Works as a whole certified by the Engineer. There shall be no maintenance period applicable to the dredging works.

Certificates and payments

(a) The Employer shall on execution of the Contract make an advance payment to the Contractor of ten per cent (10 per cent) of the Contract Price stated in the Agreement against an invoice for such amount and an approved unconditional guarantee from a Bank acceptable to the Employer in the form. Repayment of the advance payment shall commence when twenty-five per cent (25 per cent) of the value of the Works has been completed in accordance with the interim payment certificates certified by the Engineer.

(b) The Contractor shall submit to the Employer within eight (8) days after the end of each month an invoice, in a format to be approved by the Engineer and the Employer, showing the estimated contract values of the Works executed up to the end of the month, including the value of materials on Site but not incorporated in the Permanent Works in the amount and under the conditions of contract, and for taxes, duties and levies paid by the Contractor and reimbursable by the Employer.

Within thirty (30) days of receipt of the Contractor's invoice the Engineer shall issue a Certificate stating the amount properly due to the Contractor in terms of value of work executed up to the end of the month, less amounts previously certified and specifying any items in the Contractor's invoice which are not properly due to the Contractor.

(c) The Contractor shall be entitled to such a sum as the Engineer may consider proper in respect of materials intended for but not yet incorporated in the Permanent Works provided that:

    (i) the materials are in accordance with the Specifications for the Permanent Works;

    (ii) such materials have been delivered to Site, and are properly stored and protected against loss or damage or deterioration to the satisfaction of the Engineer;

    (iii) ownership shall be deemed to vest in the Employer.

(d) The Employer shall pay the Contractor the amount specified on the Certificate less a ten per cent (10 per cent) retention on the value of the Works executed excluding payments for materials on Site, within thirty (30) days of the date of issue thereof. Repayment of the advance payment shall be deducted from the balance due after the aforesaid ten per cent (10 per cent) retention.

On the issue of the Completion Certificate for the whole of the Works, fifty per cent (50 per cent) of the retention monies will be paid to the Contractor.

(e) *Final Account*

    (i) Not later than three months after the date of issue of the Maintenance Certificate, the Contractor shall submit a draft statement of Final Account and supporting documents to the Engineer showing in detail the value of the Works executed in accordance with the Contract together with all further sums which the Contractor considers to be due to him under the Contract up to the date of the Maintenance Certificate (hereinafter called the *Contractor's Draft Final Account*).

    (ii) Within three (3) months after receipt of the Contractor's Draft Final Account and of all information reasonably required for its verification, the Engineer shall determine the value of all matters to which the Contractor is entitled under the Contract.

The Engineer shall then issue to the Employer and the Contractor a statement (hereinafter called the *Engineer's Draft Final Account*) showing the final amount to which the Contractor is entitled under the Contract.

Taxes and customs duties

(a) *Foreign taxation*

The Contract Price shall include all taxes, duties, and other charges imposed outside of Federalstan on the production, manufacture, sale and transport of the construction plants, material and supplies to be used or furnished under the Contract, and on the services performed under the Contract.

(b) *Local taxes other than construction plant*
The Contractor shall be deemed to have allowed in his Tender for all duties, company income tax, income and other taxes which may be levied in Federalstan.

(c) *Business taxes*
The Contractor shall pay all company income tax levied on net income.

(d) *Income taxes on staff*
The Contractor's staff, personnel and labour will be liable to pay personal income taxes in Federalstan in respect of such of their salaries and wages as are chargeable under the laws and regulations for the time being in force, and the Contractor shall make such deductions.

(e) *Customs duties and taxes on plant*
The Contractor shall pay all taxes and customs duties imposed in Federalstan on initial importation of construction plant required for the project. The Employer will reimburse the Contractor such taxes and duties paid for the construction plant against a Bank Guarantee to be provided by the Contractor to the Employer to cover liability of taxes and customs duties with the condition that after the completion of the project the construction plant will be re-exported from Federalstan under the draw-back system imposed by the laws. The Bank Guarantee shall be in an amount equivalent to the full taxes and customs duties paid by the Contractor on the initial import. The Contractor shall in no case sell or dispose of in Federalstan any construction plant and/or spare parts on the completion of the Works and in case of default on the part of the Contractor the Guarantor on his behalf will have to pay the total value of the taxes and customs duties already reimbursed to the Contractor.

Progress reports, meetings and photographs

(a) The Contractor shall provide to the Engineer's Representative (hereinafter called *Resident Supervisor*) a regular monthly report not later than seven (7) working days after the end of the month, detailing progress on the various phases of the various parts of the Works in relation to the Contractor's Programme.

(b) Progress meetings shall be held at regular intervals but not exceeding once monthly. The meeting shall be attended by the Contractor's Project/Site Manager, Resident Supervisor of the Engineer and the Employer or his representative.

(c) Any minutes of meetings duly signed by all parties shall constitute an authorized record of matters discussed, but shall not replace any requirements in the Contract for requests for approval, instructions and decisions to be submitted in writing.

(d) Together with each progress report, the Contractor shall provide progress photographs.

1.3 Technical Specification

The following clauses are examples of the Technical Specification for the project. Because of the space limitations within the book, it is impossible to provide all the Technical Specification. Consequently it has been decided to limit the text in this section to a listing of the contents for the Specification together with section VII which is the full Specification for the concrete work for the project.

Contents for the Technical Specification

I. GENERAL
  101. Conditions of contract and specifications
  102. Temporary works
    (1) General
    (2) Location and layout
    (3) Site laboratory
    (4) Office for the Employer and the Engineer
    (5) Construction roads
    (6) Removal
  103. Soil investigation
    (1) General
    (2) Report
    (3) Technical requirements
  104. Site conditions
    (1) Location
    (2) Climate, etc.

II. GENERAL PROVISIONS
  201. Standards
  202. Test certificates

III. SITE CLEARANCE
  301. General
  302. Notice

IV. EARTHWORK
  401. Surveying
    (1) Site survey

IX. STEELWORKS

901. General

902. Steel
    (1)   Steel plates
    (2)   Steel tubes

903. Working drawings

904. Welding
    (1)   General
    (2)   Welders
    (3)   Testing
    (4)   Standards
    (5)   Paint for marine structures

X. FENDER SYSTEM

1001. General requirements
    (1)   Requirements for fender
    (2)   Standards

1002. Material
    (1)   Rubber fender
    (2)   Anchor bolts

XI. BOLLARDS

1101. General

1102. Installation, etc.

Detailed specification for section VII

VII. CONCRETE WORK

701. General

All concrete work shall be in accordance with BS 8110:1985, Part 1 and Part 2: "The Structural Use of Concrete". The Contractor shall prepare all Working Drawings, in so far as required for a proper execution of the Works, showing holes, box-outs, ducts, cast-on items and the like.

702. Cement

(1) *Type of cement*

The cement shall be sulphate-resisting Portland Cement and shall comply with BS 4027:1980.

(2) *Delivery of cement*

Cement shall be supplied with a test certificate in accordance with clause 202, and in approved bags or drums. On each bag shall be clearly indicated the initials of the type and quality of the cement and its manufacturer.

(3) *Storage at site*

The cement shall be stored in suitable dry and well-ventilated lock-up sheds. The cement sheds shall be provided with a perfectly dry floor of timber or concrete with raises at least 20 cm above existing ground level. Each consignment shall be stored separately and cement in bags shall not be stacked more than 2 m high. The Contractor shall record the cement, especially the date of dispatch, the date of delivery at site, the number of consignments, the location of storage, etc.

703. Aggregates for concrete

(1) *General requirements*

The aggregates (fine and coarse) for all grades of concrete and mortar shall comply with BS 882:1983. The aggregates shall contain no harmful material in such quantities as to give adverse effects to the strength, durability or corrosion of reinforcements. The source of all aggregates shall be approved by the Engineer.

(2) *Storage of aggregates*

The Contractor shall provide proper means of storing in such a manner that different types and grades or foreign matters cannot mix. Storage shall be capable of draining freely.

(3) *Fine aggregate*

Fine aggregate shall comply with Grading Zone 2 or 3 of BS 882:1983, and shall be obtained from natural quartz or by crushing clean granite or from freshwater river deposits. The following requirements shall also be met:

- dust in case of aggregate
- by crushing rocks: BS 812
- clay lumps: ASTM C 142
- soundness test: ASTM C 88

(4) *Coarse aggregate*

Coarse aggregate shall comprise natural gravel or natural crushed rock taken from a source approved by the Engineer. The following requirements meet:

- flakiness index: BS 812 Part 1
- crushing value: BS 812 Part 3
- clay lumps: ASTM C 142
- soundness test: ASTM C 38

704. Water

Water used for mixing and curing of concrete shall be clean, fresh water. It shall be free from oils, chemicals, vegetable matter, organic and other impurities and shall comply with BS 3148:1980.

705. Admixtures

Admixtures rates, proportions, and methods of application shall be strictly in accordance with manufacturer's instructions. Additives containing chlorides shall not be permitted for use.

706. Concrete mixes

(1) *Grade of concrete*

The following requirements shall be met with the application of concrete to jetty, prestressed concrete piles, and caisson:

- compressive strength at 28 days: $30N/mm^2$
- maximum aggregate size: 20 mm
- maximum water/cement ratio: 0.45
- minimum cement content: 360 kg/m^3

(2) *Trial mixing*

Following the Engineer's approval of the mix design, the Contractor shall carry out, in the presence of the Engineer, a trial mixing of that grade's concrete.

707. Formwork

(1) *Construction of formwork*

The formwork shall be of timber, steel or other suitable and approved material, which shall be sufficiently strong, impervious and resistant to the action of concrete.

All exposed edges of the concrete shall have a chamfer which will be specified by the Engineer. If no size is specified, the chamfer shall be 20 mm by 20 mm.

Immediately before starting concrete pouring, the interior of the forms shall be cleaned of all rubbish and the face in contact with the concrete shall be clean and treated with a composition mould oil.

(2) *Removal of formwork*

No formwork shall be removed without the approval of the Engineer, but such approval shall not relieve the Contractor of any of his obligation and responsibility.

The minimum age of the concrete when the formwork is being removed shall be as follows:

- formwork for vertical faces as caisson: 3 days
- formwork for vertical faces of jetty: 7 days
- formwork for soffit faces of jetty: 14 days

These times can be reduced with the use of rapid hardening cement, subject to the approval of the Engineer.

708. Joints

(1) *Construction joints*

Construction joints shall be so arranged as to minimize the possibility of the occurrence of initial thermal and/or drying shrinkage cracks.

If the concrete is extended above the horizontal joint on the exposed face it shall be cleaned off and washed with clean water before the next lift is placed.

Horizontal construction joints shall be covered with a 10 mm thick or freshly mixed mortar of the same sand-cement ratio as the concrete, well worked into the surface immediately before placing new concrete.

(2) *Expansion joints*

Joint-sealing compounds, joint filler and water stops shall be used in accordance with manufacturer's instructions.

Joint-sealing compounds shall be impermeable ductile materials of a suitable type for conditions of exposure in which they are to be placed, and capable of providing a durable, flexible and watertight seal by adhesion to the concrete. Two-part polysulphide-based sealants shall comply with BS 4254:1983.

Joint fillers shall be of such quality that they can be installed in position. They shall be resistant to extruding under water-compressive weathering, and shall have at least 70 per cent recovery after the application of a load to compress the filler by 50 per cent.

709. Batching, mixing and placing of concrete

(1) *Batching*

All materials for each batch of concrete shall be accurately measured by weight.

(2) *Mixing*

Concrete shall be mixed in mechanical mixers of the weigh-batch type, complying with BS 1305:1974, and equipped with an approved weight-measuring device.

The mixer drum shall be thoroughly cleaned out prior to the next mixing. Concrete which has started setting shall not be remixed either with or without additional water.

(3) *Measures to deal with high temperature*

The Contractor shall take necessary measures to ensure that the temperature of poured concrete shall be less than 30°C:

– stockpiles of each material shall be sited in shade;

– aggregate stockpiles shall be regularly and evenly sprayed with water to encourage cooling;

– mixing plant and delivery equipment shall be organized in such a way that the interval between mixing and placing is kept to a minimum;

– the mixing drum and other equipment shall be painted and kept white, wherever possible;

– immediately before placing, formwork and reinforcement shall be sprayed with cold water;

– placing during the daytime shall be avoided, wherever possible using morning and evening time.

(4) *Transportation and placing*

The Contractor shall submit to the Engineer, prior to the date of placing, a programme and schedule of concrete placing for approval, stating the start and finish time of placing and the order of placing, etc. During concreting a steel fixer and carpenters shall be in attendance. Placing during rain is not allowed without complete cover.

(5) *Compaction*

Care must be taken to ensure that compacting tools do not contact with formwork, reinforcement or any embedded fittings and to prevent the operation from transmitting any harmful vibrations to freshly placed concrete not yet hardened sufficiently.

(6) *Finishes of concrete*

Ordinary finish shall be of equivalent quality as a timber-formed concrete surface and applied to the finishes of caisson concrete.

Smooth finish shall be of equivalent quality as a steel-formed concrete surface and applied to the finishes of jetty concrete.

710. Curing of concrete

Freshly placed concrete shall be protected from sun, winds and rain, and shall be kept moist by covering with wet curing mats until it has hardened sufficiently. Unless otherwise specified, the curing shall continue till the concrete has a compressive cube strength of $10N/mm^2$. This period shall be at least seven days after placing. In no circumstances shall traffic be permitted on the concrete.

711. Reinforcement

(1) *Quality*

Steel reinforcement shall be:

"Cold-worked deformed high-yield steel bars Grade 460, complying with BS 8110".

(2) *Delivery and storage*

Reinforcement shall be supplied with a test certificate in accordance with clause 202. On the outside of all bundles shall be indicated the manufacturer, quality and quantity, the diameter, the length, etc. They shall be stored in racks clear of the ground and shall be protected from damage and oil, etc.

(3) *Working Drawings and bedding schedule*

The Contractor shall prepare all Working Drawings and bedding schedules two months before its starting, for the Engineer's checking and approval. The splices of reinforcing bars and their locations shall be in accordance with BS 4466 and shown on the Drawings.

(4) *Concrete cover to reinforcement*

The concrete cover to reinforcement shall not be less than at any place values in the following:

Location	Cover
piles	60 mm
jetty	80 mm
caisson	80 mm

(5) *Fixing*

All reinforcement shall be fixed in position by tying intersections with ample lashings of 16-gauge soft annealed iron wire, by attaching adequate concrete spacer blocks or any other approved spacers to ensure that the bars are not displaced during concreting.

1.4 Bill of Quantities

The following is the Bill of Quantities for the project.

GENERAL PREAMBLE

(1) The Contract Documents contain provisions and requirements essential to the Bill of Quantities and the Bill of Quantities shall be read in conjunction with these Documents. Tenderers shall be deemed to have examined fully the Contract Documents and to have satisfied

themselves with the requirements of the Works to be performed under the Contract.

(2) The quantities set against the various items herein are estimates of the quantities included in the Contract, and are not taken as a warranty that the exact quantities scheduled in the Bill will be carried out. The work actually carried out will be measured by the Engineer and paid for at the Scheduled Rate set forth in the Bill of Quantities irrespective of whether the final quantities increase or decrease.

(3) The price and the Schedule Rates to be entered in the Bill of Quantities shall be the full inclusive value of the work, including all incidental and contingent costs, profits, overheads, allowances, taxes, duties, insurance, risk, and liabilities of every kind implied in the Contract Documents.

Such cost shall be deemed to include purchase of material, inspection, testing, packing, storing, delivery, installation, and construction costs and depreciation costs of machines as well as the administrative and management costs inclusive of the temporary work yard, offices, accommodation, water and electricity supply for the construction works and living, and other necessary expenses incurred as Site Overhead Costs.

(4) The item "Descriptions" in the Bill are only brief and explanatory descriptions, and detailed directions of workmanship and materials given in the Specifications are not necessarily repeated in the Bill of Quantities.

(5) The actual sums to be paid to the Contractor shall be determined by measuring the works actually done in accordance with the terms of Contract and at the Scheduled Rates or Lump-Sum Price. The items priced on a lump-sum basis shall be paid at the same price, irrespective of the actual increase or decrease of the quantity of the Works.

(6) The Bill of Quantities is to be priced in Federalstani Nu. When tendering, the Contractor shall enter the percentage of the amounts against each section of the Bill of Quantities that are required to be paid in foreign currencies. The exchange rates shall be established in accordance with clause (k)(ii) of the Instructions to Tenderers.

(7) In case of conflict or discrepancies, errors or omissions among the Contract Documents the matter shall be informed by the Contractor to the Engineer immediately for clarification. Any work affected by such conflicts, discrepancies, errors, omissions shall be at Contractor's risk. Anything shown on the Drawings and not listed in the Bill of Quantities or vice versa shall have the same effect as if shown in both.

The list of bill items is divided into the following sections:

Bill No. 1   General items
Bill No. 2   Site preparation
Bill No. 3   Earthworks
Bill No. 4   Unloading jetty
Bill No. 5   Breakwater

A bill summary sheet is also provided.

These lists of bill items and the bill summary sheet are given in figures 1 to 8 inclusive.

Figure 1. The Bill of Quantities (sheet 1 of 7)

1. General Items   Bill No. 1

Item No.	Description	Unit	Quantity	Schedule rate	Amount (Nu)
1.1	Mobilization of equipment to the Site and demobilization	L.S.	1		
1.2	Topographic survey				
1.2.1	Survey on land	L.S.	1		
1.2.2	Sounding offshore	L.S.	1		
1.3	Soil investigation				
1.3.1	Onshore boring, including samples and field test	m	50		
1.3.2	Offshore boring, including temporary staging	m	100		
1.3.3	Laboratory test on the soil samples, including preparation of reports	L.S.	1		
1.4	Provide and maintain the following for exclusive use of the Engineers				
1.4.1	Offices as specified in clause 102 (4)	L.S.	1		
1.4.2	New vehicles for their transportation	Nos.	5		
1.4.3	New generators (300 KVA)	Nos.	2		
1.4.4	Potable water, oil and their storage tanks	L.S.	1		
1.4.5	Detached houses for accommodation	Nos.	5		

BILL TOTAL

Figure 2. The Bill of Quantities (sheet 2 of 7)

2. Site preparation   Bill No. 2

Item No.	Description	Unit	Quantity	Schedule rate	Amount (Nu)
2.1	*Site clearance*				
2.1.1	General site clearance	L.S.	1		
2.1.2	Disposal area	L.S.	1		
2.2	*Preparation of Site*				
2.2.1	Site office with gate and fencing	L.S.	1		
2.2.2	Site laboratory for tests of soil, cement, water, etc. with equipment and staffing	L.S.	1		
2.2.3	Warehouses for all materials used for the permanent works	L.S.	1		
2.2.4	Desalination plant for potable water for living and construction purposes	L.S.	1		
2.2.5	Temporary loading jetty and gantry crane	L.S.	1		
2.2.6	Temporary survey jetty and platform	L.S.	1		

BILL TOTAL

Figure 3.   The Bill of Quantities (sheet 3 of 7)

3.   Earthworks   Bill No. 3

Item No.	Description	Unit	Quantity	Schedule rate	Amount (Nu)
3.1	*Dredging*				
3.1.1	Dredging as specified on the Drawings, including disposal of excavated soils	m³	3 000 000		
3.1.2	Echo-sounding after the dredging work	L.S.	1		
3.2	*Reclamation*				
3.2.1	Fill by hydraulic means	m³	70 000		
3.2.2	Fill in the dry	m³	30 000		
3.2.3	Levelling and topping by sand-gravel mixture	m²	52 500		
3.3	*Slope protection*				
3.3.1	Supply, place and shape armour rocks 200–300 kg	m³	70 000		
3.3.2	Supply, place in position filter fabric between rocks and the ground	m²	35 000		
				BILL TOTAL	

Figure 4.   The Bill of Quantities (sheet 4 of 7)

4.   Unloading jetty

Bill No. 4

Item No.	Description	Unit	Quantity	Schedule rate	Amount (Nu)
4.1	*Piling*				
4.1.1	Supply of prestressed concrete piles Ø 750 mm, 125 mm thick, 20 m length	Nos.	96		
4.1.2	Supply of prestressed concrete piles Ø 900 mm, 150 mm thick, 26.4 m length	Nos.	24		
4.1.3	Supply of steel pipe piles Ø 300 mm, 9 mm thick, 18 m length with welding for joints	Nos.	42		
4.1.4	Handle and drive vertical piles (Ø 750 mm x 125 t)	Nos.	96		
4.1.5	Handle and drive batter piles (Ø 900 mm x 150 t)	Nos.	24		
4.1.6	Handle and drive fender piles (Ø 300 mm x 9t)	Nos.	42		
4.1.7	Test piling Ø 750 mm	Nos.	2		
4.1.8	Test piling Ø 900 mm	Nos.	2		
4.1.9	Static loading test with max. loading of 250 ton, including temporary platform and reaction piles	Nos.	4		

BILL TOTAL

Figure 5. The Bill of Quantities (sheet 5 of 7)

4. Unloading jetty (*cont.*)

Item No.	Description	Unit	Quantity	Schedule rate	Amount (Nu)
4.2	*Concrete work*				
	Chipping				
4.2.1	Chipping and disposal of pile top for ∅ 750 mm piles	Nos.	96		
4.2.2	Chipping and disposal of pile top for ∅ 900 mm piles	Nos.	24		
4.2.3	Cutting and disposal of pile top for ∅ 300 mm piles	Nos.	42		
	Pile caps				
4.2.4	Filling concrete with reinforcement cage for ∅ 750 mm piles	Nos.	96		
4.2.5	Filling concrete with reinforcement cages for ∅ 900 mm piles	Nos.	24		
	Deck slab concrete				
4.2.6	Deck slab concrete	m^3	4 500		
4.2.7	Curb (150 mm x 150 mm)	m^3	7		
	Reinforcement				
4.2.8	Reinforcement Grade 460 for 4.2.6 above (a) up to 10 mm dia. (b) up to 19 mm dia. (c) up to 29 mm dia.	t t t	27 108 135		
				BILL TOTAL	

Figure 6.   The Bill of Quantities (sheet 6 of 7)

4.   Unloading jetty (*cont.*)   Bill No. 4

Item No.	Description	Unit	Quantity	Schedule rate	Amount (Nu)
	Formwork				
4.2.9	Setting and removal of soffit formwork and falsework	m²	3 000		
4.2.10	Setting and removal of side formwork	m²	450		
4.2.11	Setting and removal of stop-end formwork	m²	180		
4.3	*Fender system*				
4.3.1	Supply and installation of cell-type rubber fender	Nos.	42		
4.3.2	Supply and installation of rubbing strip	Nos.	14		
4.3.3	Anti-corrosion coating or painting upon the completion	L.S.	1		
4.4	*Miscellaneous*				
4.4.1	Supply and installation of galvanized steel tube ladder	Nos.	2		
4.4.2	Supply and installation of 50 t bollard	Nos.	12		
4.4.3	Expansion joint (a) galvanized steel angle 150 x 150 × 10	m	80		
	(b) buffer plate	Nos.	4		

BILL TOTAL

Figure 7. The Bill of Quantities (sheet 7 of 7)

5. Breakwater

Bill No. 5

Item No.	Description	Unit	Quantity	Schedule rate	Amount (Nu)
5.1	*Caisson mound*				
5.1.1	Supply, place and shape rubble bedding stone (50-100 kg/pc)	m^3	62 400		
5.1.2	Supply, place and shape armour rocks (200-300 kg/pc)	m^2	37 200		
5.1.3	Supply and place in position filter fabric	m^3	6 000		
5.1.4	Inspection by sounding of mound formation	L.S.	1		
5.2	*Concrete work*				
	In situ concrete				
5.2.1	Caisson concrete	m^3	17 960		
5.2.2	Topping concrete	m^3	6 675		
5.2.3	Non-shrinkage grout for gaps between caissons	Nos.	59		
	Reinforcement				
5.2.4	Grade 460/425 for 5.2.1 (a) up to 10 mm dia. (b) up to 19 mm dia. (c) up to 29 mm dia.	t t t	18 1 257 521		
5.2.5	Grade 460/425 for 5.2.2 (a) up to 10 mm dia. (b) up to 19 mm dia. (c) up to 29 mm dia.	t t t	33 234 67		

BILL TOTAL

Figure 8. The bill summary sheet

	GRAND SUMMARY		Total (Nu)
Bill No. 1	Sheet 1		
	(Foreign currency element	%)	
Bill No. 2	Sheet 2		
	(Foreign currency element	%)	
Bill No. 3	Sheet 3		
	(Foreign currency element	%)	
Bill No. 4	Sheet 4		
	(Foreign currency element	%)	
	Sheet 5		
	(Foreign currency element	%)	
	Sheet 6		
	(Foreign currency element	%)	
Bill No. 5	Sheet 7		
	(Foreign currency element	%)	

Signed: _____

On behalf of
Tenderer: _____

Address: _____

_____

_____

_____

TOTAL TENDER SUM

1.5   Tender Drawings

The following figures 9 to 13 inclusive are examples of the Tender Drawings supplied by the Engineer. These are as follows:

Drawing No. 001   The overall plan of the works
Drawing No. 002   The jetty plan and elevation
Drawing No. 003   A typical section of the jetty
Drawing No. 004   A typical section of the breakwater
Drawing No. 005   Details of the caisson

Figure 9. Drawing 001. The overall plan of the works

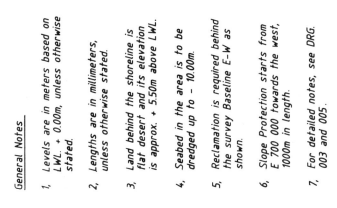

General Notes

1, Levels are in meters based on LWL. + 0.00m, unless otherwise stated.

2, Lengths are in millimeters, unless otherwise stated.

3, Land behind the shoreline is flat desert and its elevation is approx. + 5.50m above LWL.

4, Seabed in the area is to be dredged up to - 10.00m.

5, Reclamation is required behind the survey Baseline E-W as shown.

6, Slope Protection starts from E 700 000 towards the west, 1000m in length.

7, For detailed notes, see DRG. 003 and 005.

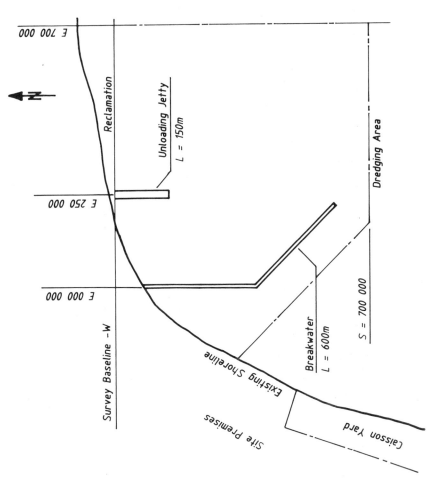

E 700 000

Reclamation

Unloading Jetty
L = 150m

E 250 000

E 000 000

Survey Baseline -W

Existing Shoreline

Site Premises

Breakwater
L = 600m

Dredging Area

S = 700 000

Caisson Yard

Figure 10. Drawing 002. The jetty plan and elevation

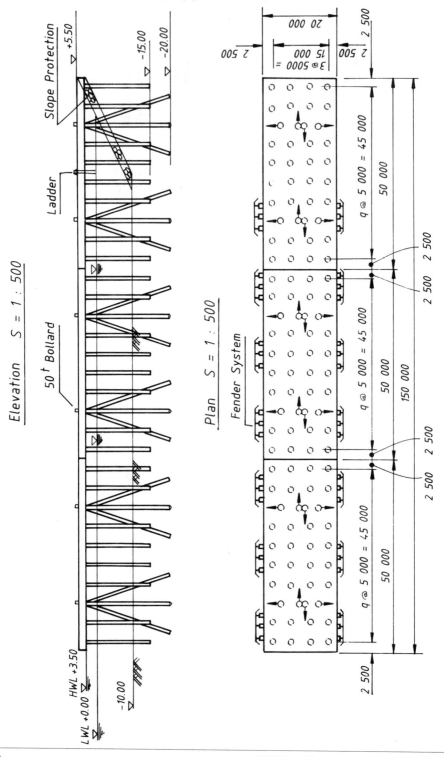

Figure 11.   Drawing 003.   A typical section of the jetty

Notes

1. The jetty is designed for a general cargo up to 20 000DWT. with its berthing velocity of 0.20 m/sec.

2. The jetty is earthquake-resistant

3. Foundation Piles : PC pile
    - vertical Ø750 * 125
      tip elev. = - 15.0m
    - batter Ø900 * 150 (3 : 1)
      tip elev. = - 20.0m

4. Fender piles : Steel pipe pile
    Ø300 * 9 (50 : 1)
    tip elev. = - 12.0m

5. Deck slab is of insitu reinforced concrete in accordance with the following requirements:
    - concrete : 300 kg/cm
    - cement : sulphate resistant cement to BS 4027
    - rebar : Grade 460/425 to BS 4461

6. Vertical concrete surface contacting soils shall be protected by painting asphalt-bitumenous material.
    - 50 bollard
    - cell rubber fender
    - ladder with galvanised surface
    - robbing strip with structural steel and rubber

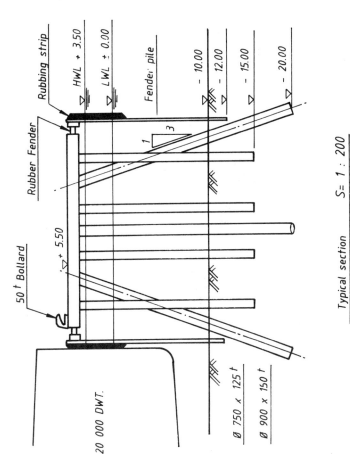

Rubbing strip

HWL + 3.50

LWL ± 0.00

Fender pile

- 10.00

- 12.00

- 15.00

- 20.00

Rubber Fender

+ 5.50

50 t Bollard

20 000 DWT.

Ø 750 x 125 t

Ø 900 x 150 t

Typical section   S= 1 : 200

Figure 12. Drawing 004. A typical section of the breakwater

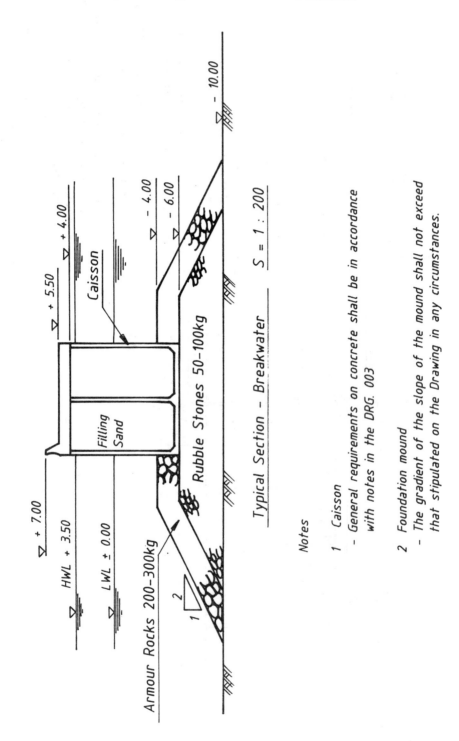

Typical Section – Breakwater   S = 1 : 200

Notes

1 Caisson
  - General requirements on concrete shall be in accordance
    with notes in the DRG. 003

2 Foundation mound
  - The gradient of the slope of the mound shall not exceed
    that stipulated on the Drawing in any circumstances.

Figure 13.   Drawing 005.   Details of the caisson

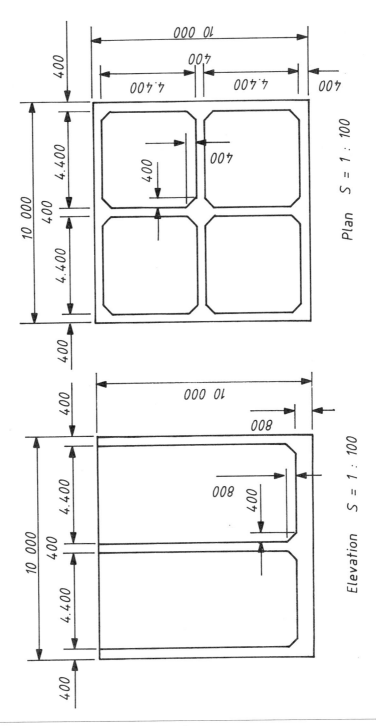

1.6 Form of Tender

Federalstan Harbour Authority,
Contract Road, Capital City,
FEDERALSTAN.

GENTLEMEN

Having examined the Tender Documents for the above Contract, including Drawings, Conditions of Contract, Specifications and Bill of Quantities for the construction of the above-mentioned Works, the receipt of which is hereby acknowledged, we offer to construct and maintain the whole of the Works as described in, and in accordance with, the said Tender Documents including Addenda Nos._____, for the sum of _____ Nu.

We undertake, if our Tender is accepted, to commence the Works within _____ days and to complete and deliver the Works in accordance with the Contract within _____ days calculated from the date of commencement of the Works and in accordance with the Time Schedule.

If our Tender is accepted we will provide the Performance Bond in the sum of _____ , equal to 10 per cent of the Contract Price, for the due performance of the Contract.

We agree to abide by this Tender for the period of one hundred and twenty (120) days from the date fixed for Tender Closing pursuant to clause (n) of the Instructions to Tenderers and it shall remain binding on us and may be accepted at any time before the expiration of that period.

We attach the Appendices to the Tender Form, duly completed and signed.

Until a formal Contract is prepared and executed, this Tender, together with your written acceptance thereof by your notification of Award, shall constitute a binding contract between us.

We understand that you are not bound to accept the lowest or any Tender that you may receive.

We are, Gentlemen,

Yours faithfully,

Signature: _____

Address: _____

_____

_____

Date: _____

Appendix to the Form of Tender

Details in this Appendix are cross-linked to the Conditions of Contract by the clause numbers shown:

	Clause	
Amount of Bond	10	10 per cent
Minimum amount of third party insurance	23	60 000 000 Nu
Period of commencement	41	Within thirty (30) days after receipt of Engineer's order to commence
Time for completion	43	540 days calculated from the last day of the period of commencement
Limit of liquidated damages	47	10 per cent of Contract Price
Period of maintenance	47	365 days
Percentage of retention	60	10 per cent
Minimum amount of interim certificate	60	300 000 Nu
Time within which payments to be made after issue of Certificates	60	30 days

Note: This Form of Tender is adapted from the FIDIC Conditions of Contract.

1.7 Form of Bid Bond

Whereas _____ (hereinafter called Tenderer) has submitted its Tender dated _____ for the construction of a new cargo jetty and breakwater (hereinafter called the Tender) _____ know all men by these present that WE the Bank of _____ _____ having our registered office at _____ (hereinafter called the Bank) are bound to Federalstan Harbour Authority (hereinafter called the Employer) in the sum of _____ Nu for which payment to be made to the said Employer, the Bank binds itself, its successors and assigns by these present. Sealed with the Common Seal of the said Bank this _____ day of _____ 19__ .

The CONDITIONS of this obligation are:

1. If the Tenderer withdraws its Tender during the period of Tender validity specified by the Tenderer on the Tender form; or

2. If the Tenderer, having been notified of the acceptance of its Tender by the Employer during the period of its Tender validity;

   (a) fails or refuses to execute the Contract Form when requested; or

   (b) fails or refuses to furnish the Performance Bond, in accordance with the Instructions to Tenderers.

We undertake to pay to the Employer up to the above amount according to and upon receipt of its first written demand, without the Employer having to substantiate its demand, provided that in its demand the Employer shall note that the amount claimed by it is due to it owing to the occurrence of one or both of the two above-stated conditions, specifying the occurred condition or conditions.

This guarantee will remain in force up to and including thirty (30) days after the period of Tender validity, and any demand in respect thereof should reach the Bank not later than such date.

Name of Bank: _____

Signature of authorized
Representative of Bank: _____

Signature of Witness: _____

Name of Witness: _____

Address: _____

_____

_____

_____

Note: This Form of Bid Bond is adapted from the FIDIC Conditions of Contract.

1.8   Form of Agreement

THIS AGREEMENT is made on the _____ day of _____ nineteen hundred and _____.

Between the Federalstan Harbour Authority of the one part and _____ (hereinafter called the Contractor) of the other part.

Whereas the Employer is desirous that certain works should be executed, viz. the construction of a new cargo jetty and breakwater, and has accepted a Tender by the Contractor for the execution, completion and maintenance of such works. NOW THIS AGREEMENT WITNESSETH as follows:

1. In this Agreement words and expressions shall have the same meanings as are respectively assigned to them in the Contract hereinafter referred to.

2. The Contract shall consist of this Form of Agreement, and the following documents, all of which by this reference are incorporated herein and made part hereof:

   (a) the Letter of Acceptance;

   (b) the Tender and Appendix to Tender;

   (c) the Technical Specifications;

   (d) the priced Bill of Quantities;

   (e) the Drawings;

   (f) the Work Schedule;

   (g) the Schedules of Supplementary Information;

   (h) the Conditions of Contract;

   (i) the Instructions to Tenderers.

3. The aforesaid documents shall be taken as complementary and mutually explanatory to one another, but in the case of ambiguities or discrepancies precedence shall be taken in the order set out above.

4. In consideration of the payments to be made by the Employer to the Contractor as hereinafter mentioned the Contractor hereby covenants with the Employer to execute, complete and maintain the Works in conformity in all respects with the provisions of the Contract.

5. The Employer hereby covenants to pay the Contractor in consideration of the execution, completion and maintenance of the Works the Contract Price at the times and in the manner prescribed by the Contract.

In WITNESS whereof the parties hereto have caused their respective common seals to be hereunto affixed (or have hereunto set their respective hands and seals) the day and year first above written.

Signed, sealed and delivered

On behalf of the Employer   On behalf of the Contractor

By: _____   By: _____

Name: _____   Name: _____

Capacity: _____   Capacity: _____

In the presence of: In the presence of:

_____   _____

Name: _____	Name: _____
Address: _____	Address: _____
_____	_____
_____	_____
_____	_____

Note: This Form of Agreement is adapted from the FIDIC Conditions of Contract.

1.9 Scope of the Works

The Works, which shall be performed in strict accordance with the provisions of the Agreement, the Drawings and the Specifications, comprise the execution, completion and maintenance of a New Cargo Jetty and Breakwater, as outlined hereunder:

- dredging and reclamation with compaction of fill material;
- slope protection;
- construction of reinforced concrete jetty;
- supply and installation of bollards, fenders and other ancillary items;
- construction of caisson-type breakwater with foundation works, filling sand, and *in situ* concrete on the top of caissons;
- provision of all as-built drawings and other documentation as stipulated in the Specification.

1.10 Instructions relating to the programme for the Works

1. *Completion*

The Contractor shall complete the whole of the Works within five hundred and forty (540) days calculated from the last day of the Period for Commencement.

The date of completion for the purposes hereof shall mean the date of substantial completion stated in the Certificate of Completion issued by the Engineer, pursuant to clause 48 of the Conditions of Contract.

2. *Progress*

The attention of the Contractor is drawn to the provisions of clause 46 of the Conditions of Contract regarding progress of the Works.

3. *Liquidated Damages*

Liquidated Damages levied pursuant to clause 47 of the Conditions of Contract shall be applied from the date stated in section B.1 above.

1.11 Form of Performance Bond

To: Federalstan Harbour Board,
     Contract Road, Capital City,
     Federalstan.

Whereas _____ (hereinafter called the "Contractor") has undertaken to construct and complete the "New Cargo Jetty and Breakwater (EDB Loan 1000-91)" in pursuance of Contract No. _____ dated _____ hereinafter called "the Contract";

And whereas it has been stipulated by you in the Contract that the Contractor shall provide you with a Bank Guarantee by a recognized Bank for the sum specified therein as security for compliance with the Contractor's performance obligations in accordance with the Contract.

And whereas we have agreed to give the Contractor a Guarantee;

Therefore we hereby affirm that we are Guarantors and responsible to you, on behalf of the Contractor, up to a total _____, and we undertake to pay you, upon your first written demand declaring the Contractor to be in default under the Contract, and without cavil or argument, any sum or sums as specified by you, within the limit of the amount as aforesaid, without needing to prove or to show grounds or reasons for your demand or the sum specified.

THIS Guarantee is valid until the _____ day of _____ 19__.

Name of Guarantor: _____

Title: _____

Date: _____

Address: _____

Note: This Form of Performance Bond is adapted from the FIDIC Conditions of Contract.

1.12 Form of Guarantee for Advance Payment

GENTLEMEN

In accordance with the provision of the Conditions of Contract, Messrs. _____, the Contractor under the terms of Contract has to deposit with NEW HARBOUR PROJECT an Advance Payment Security to guarantee his proper and faithful performance of the Contract in the amount of _____ Nu, we, the _____ as instructed by the Contractor, agree unconditionally and irrevocably to guarantee as primary obligator and not surety merely, the payment to

Federalstan Harbour Authority on its first demand without whatsoever right of objection on our part and without its first claim to the Contractor, in the amount of not exceeding _____ Nu in the event that obligations expressed in the above-stated Contract have not been fulfilled by the Contractor giving the right of claim to the Employer for recovery of the advance payment from the Contractor under the Contract.

Signed by the said Guarantor: _____

In the presence of: _____

Signed by: _____

For and on behalf of the Principal
in the presence of:

_____

Note: This Form of Guarantee is adapted from the FIDIC Conditions of Contract.

AN APPRAISAL OF THE CONTRACT DOCUMENTS AND THE PREPARATION OF THE PRE-TENDER PROGRAMME

2

2.1 An appraisal of the Contract Documents

The following analysis was made of the Contract Documents by the Contractor's Chief Estimator when considering whether or not to proceed with the Tender.

Instructions to Tenderers

Source of Funds

The project is to be funded by a loan "received" from the Euroasian Development Bank, in "various currencies". It would appear that monies have been allocated for the project. A review of the "Guidelines for Procurement under EDB Loans" would be useful plus additional confirmation of funding. Has our representative in Federalstan confirmed the intention of the Harbour Authority to proceed with this project?

Eligibility requirements

As a registered Unified Republic company we are eligible to be the main Contractors for the Works. A check on the list of eligible source countries includes a sufficient range of countries for us to ensure appropriate subcontractors and materials suppliers and to prepare a realistic Tender.

Scope of the work

The Employer is the Federalstan Harbour Authority, an internationally recognized government authority. Given the scope of the works:

- site clearance;
- new unloading jetty (150 m × 20 m);
- breakwater (caisson type 600 m long);
- dredging (3,000,000 m³);
- reclamation (100,000 m³); and
- slope protection (1,000 m),

the time-span allowed is short. No intended date for commencement is provided. The availability of specialist marine equipment and speedy mobilization of the site will be critical.

Pre-tender site visit

A site visit is essential. Given the tender submission date, it is suggested that this visit is organized immediately.

Preparation of the Tender

English is the accepted language for all correspondence and the submission of the Tender.

Submission of the Tender

It is required to submit:
- Form of Tender;
- Bill of Quantities;
- Schedules; and
- Bid Bond.

It is not clear what is required in the way of Schedules. Clarification of this point is required.

Tender prices

Not all items need to be priced. Unit rates and lump-sum prices are allowable.

The currencies for the Tender and for payment

Unit rates and prices to be quoted in Federalstani Nu. A breakdown of all other foreign currency payments is required. Exchange rates to be used at the time of Tender are those thirty days (30) prior to submission. No statement is given as to the exchange rates to be used when payment for completed works is authorized.

Alternative designs

Given our company's expertise in marine works, an alternative design may be appropriate. However, design staff availability and the short tender period may be prohibitive.

Note: Additional Bid Bond requirements required for alternative designs.

The validity of the Tender

Tender valid for 120 days, plus extension if required. Check cash flows and currency implications.

The Conditions of Contract

The following points are noted:

- all documentation will be in the English language;
- Contract to be construed in accordance with the laws of the Republic of Federalstan;
- copies of all working drawings required 12 weeks before commencement. No clear procedure for approval stated;
- minimum Contractors Third Party Insurance 60 million Nu per occurrence;
- no night or Friday work save tidal marine works;
- a maximum of 10 per cent liquidated damages;
- 12 months' maintenance.

Advance payment of 10 per cent available on execution of the contract, but an Acceptance Guarantee required. Repayment of initial payment in stages after 25 per cent of the value of the works completed. Cash flow implications need to be considered. Retention monies at the rate of 10 per cent will be withheld, 50 per cent retention repayable on issue of the Completion Certificate.

All local taxes including Business Tax will be levied and the tender must accommodate all such costs. Importation tax on all construction plant imported into the country will be reimbursed provided the plant is exported after completion of the project. It is prohibited to sell any construction plant or parts within Federalstan.

The technical specification

Concrete work

All standards quoted within the specification are British Standards with which the company is familiar. Special clauses relate to the storage of materials for concrete, and specific measures are quoted for the pouring of concrete given the high ambient temperature. These will affect the storage conditions on site and working procedures.

Bill of Quantities

The Bill of Quantities provided is restricted to some 50 items. This will facilitate an estimate based on an operational basis.

Form of Tender, Form of Bid Bond,
Form of Agreement, Form of Performance Bond
and Form of Guarantee for Advance Payment

All these documents appear to be standard. Copies have been forwarded to the Legal Department for review when formally instructed.

An analysis of other factors and recommendations

As a company, we have for some time been actively seeking work in this area. Our current workload is low and we have suitable staff and plant resources available to undertake the work. The client is an internationally recognized government authority. Provided our agent and financial advisers are in a position to confirm that funding has been allocated to the project we should submit a tender with the view to securing the work.

2.2   The site organization

The site staff organization will be structured as shown in figure 14.

Site office, storage, working yard, plant yard and warehouses necessary to execute the project will be established in the site premises area shown in figure 15.

The following facilities are neither specified nor shown on the above drawings:

- generators as power sources;
- power line, water pipeline;
- lighting;
- oil and fuel storage tanks; and
- drainage system.

The layout of these offices was prepared based on the following:

- a southerly wind is predominant throughout the year (dust from the crushing plant does not cause problems inside the site);
- the convenience of transportation;
- accommodation will be established a suitable distance from the site.

2.3   The construction method statement

The method statement is divided into a number of sections:

- dredging;
- reclamation;
- jetty construction; and
- the caisson for the breakwater.

Each of these sections is now described in detail.

Figure 14.   The site staff organization

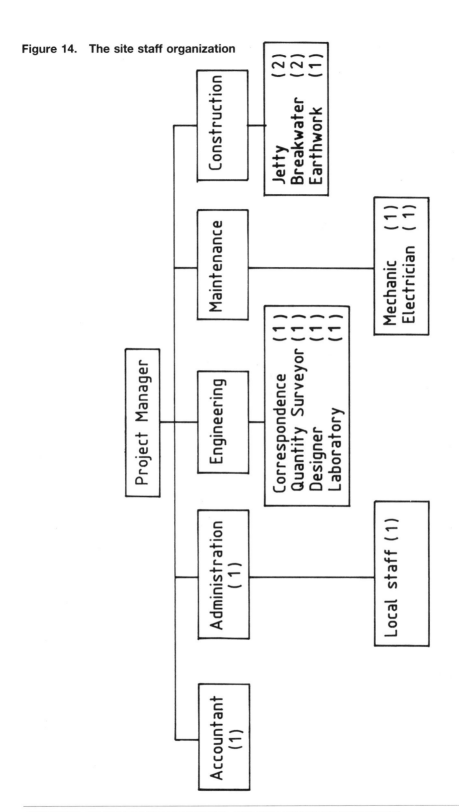

Figure 15.   The layout of the site premises

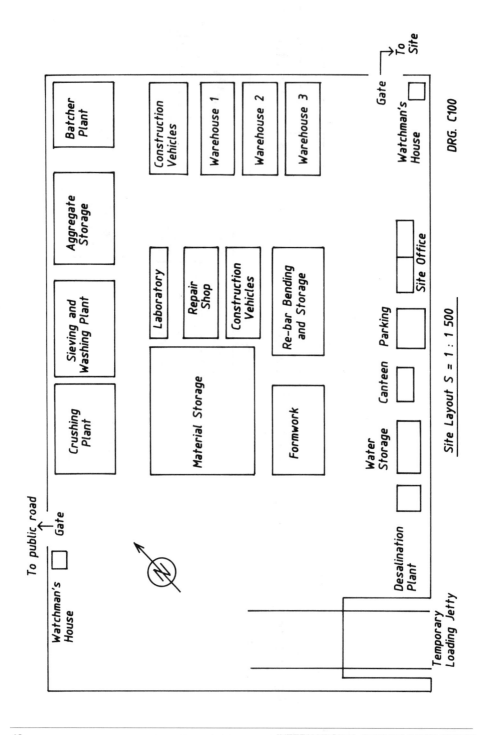

2.3.1 Dredging

(a)   The following dredger shall be used:

Length of dredger:	80 m
Width of dredger:	20 m
Depth:	5 m
Draft:	4 m
Maximum depth of dredging:	30 m
Dredging capacity:	1 500 m³ /hour
Maximum distance of discharge:	8 000 m
Power source:	steam turbine
Pump power:	1000 PS

(b)   The dredger, floater and discharging pipelines shall be assembled in the site, taking approximately 14 days. The discharging pipelines shall at first be extended to the reclamation area and the filling below +3.70 m in the area shall be achieved. Upon completion of the hydraulic filling, the pipe will be connected with another pipeline and extended to the designated disposal area.

(c)   The dredging work shall be carried out on a basis of 24-hour full operation with three-shift system.

(d)   Oil, fuel and any other necessary consumptions shall be supplied from the onshore site periodically or on request.

(e)   Taking into account the accumulation of soils during the Contract period and the subsequent change of the dredged level at the time of Completion Certificate, 50 cm of excess dredging shall be undertaken, i.e., the seabed shall be dredged up to −10.5 m.

An outline drawing of the dredger arrangement to be used for the project is shown in figure 16.

2.3.2 Reclamation

Reclamation work shall be divided into the following two categories of filling: (a) hydraulic filling; and (b) filling in the dry.

(a)   Hydraulic filling is defined as fill achieved underwater or in a wet condition, i.e., in this instance defined as a filling work below +3.70 m. Fill materials are the soils excavated from the existing seabed by the above-mentioned dredging work, and filled in directly from the discharging pipeline. The formation shall be made by divers and an underwater bulldozer equipped with slope-finishing attachment. (See figure 17.)

(b)   Upon completion of the hydraulic filling, the next stage of reclamation should be made in the dry, i.e., filling works from the land side (see figure 18).

Figure 16. An outline drawing of the dredger arrangements

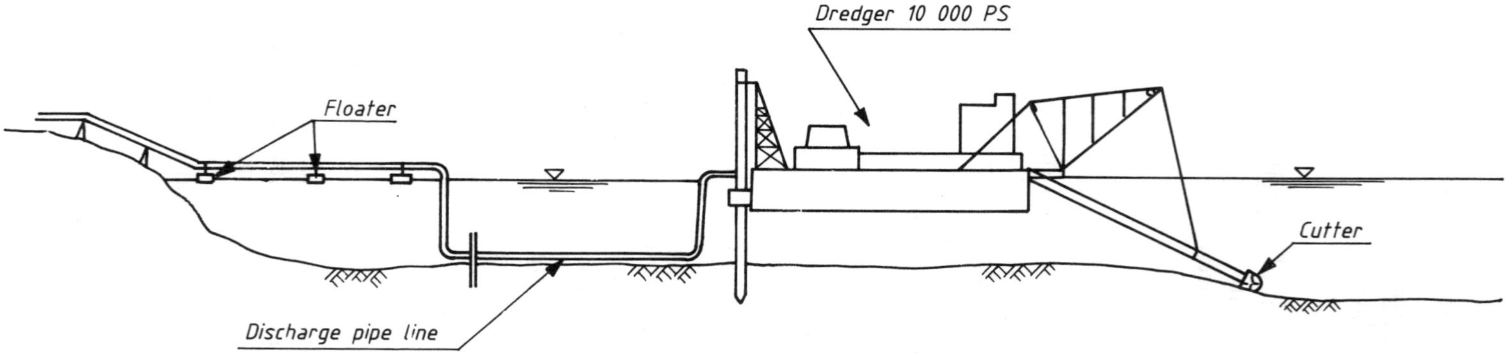

Cutter

Dredger 10 000 PS

Floater

Discharge pipe line

Figure 17. Formation of the slope for hydraulic filling

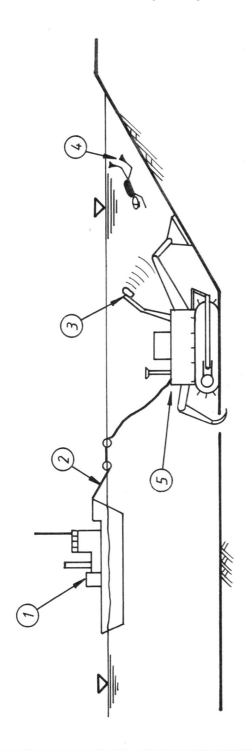

1, Commander boat with generator (175 KVA)
2, Power cable.
3, Lighting.
4, Diver.
5, Under-water bull-dozer equipped with slope-finishing attachment.

Figure 18.   Filling in the dry (above hydraulic filling)

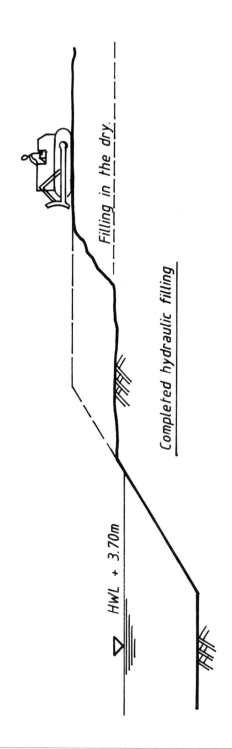

Filling materials shall be obtained from an approved borrow pit, and transported, spread, filled, shaped and compacted by the following construction plant:

Items	Nos.
Dumpers	5 (by material suppliers)
Bulldozer (D5)	3
Back-hoe (1.5 m³)	1
Grader (4.5 m)	2
Tyre-roller (11 t)	2

For detailed procedure, attention must be drawn to the clause 402 of the Technical Specification.

After compacting the whole area, soil strength and bearing capacity shall be field tested at random for confirmation and approval of the Engineer. Points for the testing shall be chosen by the Engineer during the filling and the test shall be carried out.

2.3.3 Jetty construction

(a) *Preparation for pile driving*

Prior to the commencement of pile driving, the seabed in the piling area shall be dredged or reclaimed, and shaped as specified on the Drawings.

An additional 50 cm of excess dredging will be performed because, after the completion of pile driving, it will be very difficult and expensive to dredge the area again.

(b) *Pile driving*

(i) Construction plant

The major construction plant employed in the pile driving shall be as follows:

– piling barge with steam hammer:	1 No.
– crane barge:	1 No.
– pontoon (500 t):	2 Nos.
– tugboat (1500 PS):	1 No.
– anchor boat (5 t):	1 No.
– traffic boat:	2 Nos.
– steam hammer:	1 No.
– generator:	1 No.

An outline drawing of the piling barge is shown in figure 19.

Figure 19. An outline drawing of the piling barge

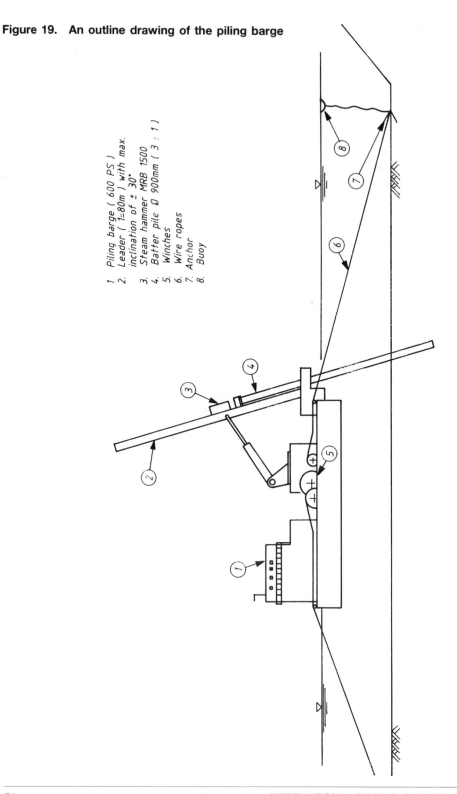

1. Piling barge (600 PS)
2. Leader (1=80m) with max. inclination of ± 30°
3. Steam hammer MRB 1500
4. Batter pile Ø 900mm (3 : 1)
5. Winches
6. Wire ropes
7. Anchor
8. Buoy

(ii)   Transportation of piles

Piles shall be stored within the working area of gantry crane, and shifted on a pontoon moored at the temporary loading jetty. They will be towed by tugboat to the piling barge. The surface of the pile shall clearly be marked with paint at suitable intervals (the top 2 m of a pile shall be at 10 cm intervals, and the rest 1 m intervals).

(iii)   Piling

After locating the pile in the leader, the pile will be positioned accurately by survey from two directions. After securing the barge in the correct position by tightening anchor ropes, pile driving will be commenced.

During the driving operation, surveyors shall continuously observe the location and verticality or inclination of the pile. If amendment of the pile's location or inclination is necessary, the barge captain will be immediately informed, driving stopped and the necessary corrections shall be made. In the case of the sloping piles great attention must be paid to the inclination because of continuous change of tidal level and subsequent bend of the axis of the pile.

(iv)   After piling

On completion of the driving, the piling barge will be shifted to the next position.

The sloping driven piles shall not be left unsupported. They will be supported with ropes or by another suitable method and kept in an accurate inclination in order to prevent unfavourable stresses.

(c)   *Slope protection within the jetty*

After finishing piling in the sloped area and prior to the commencement of falsework for the concrete deck slab of the jetty, slope protection within the width of the jetty shall be started. The placing of these rocks will be done so as not to damage the piles already driven.

(d)   *Deck slab*

(i)   Falsework

Steel brackets shall be set at all vertical piles and then steel I-beams shall be erected on the brackets as shown in figure 20.

(ii)   Chipping of pile top

Sufficient safe scaffolding and working space will be provided and the top of the piles shall be chipped off with concrete breakers without damaging the reinforcement of the piles. Debris will not be thrown into the sea. All debris will be transported to the approved disposal area.

Figure 20.   A typical section of the falsework

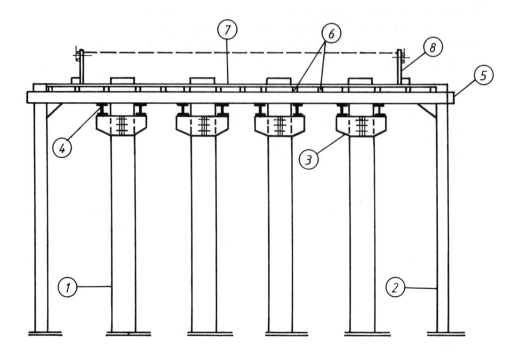

1.   Permanent p.c. pile Ø 750mm

2.   Temporary steel H-pile (300H)

3.   Steel bracket

4.   Longitudinal H-beam (400H)

5.   Transversal H-beam (300H)

6.   Processed timber 100 * 100 as secondary

7.   Soffit form (plywood t = 12mm)

8.   Vertical side form

INTERNATIONAL BIDDING CASE STUDY

(iii)   Pile cap

After completion of soffit formwork, construction of the pile caps shall be started. The placing of concrete will be by concrete bucket from the crane barge.

(iv)   Concreting

After the fixing of the reinforcement, vertical formwork and other associated support work concrete will be placed to the deck slab. Concrete will be poured from the land or from the previously completed sections of the jetty provided the concrete already placed has reached sufficient strength. The placing of the concrete to the deck slab will be made using the concrete pump.

(v)   Finishes, etc.

Concrete finishing, curing and the removal of forms shall be done as specified in clause 709(6), 719 and 707(2) respectively.

(e)   *Fender system*

Fender piles shall be driven offshore by the crane barge (50t), with the use of a vibration hammer (45 kW) and template. Then cylindrical rubber fenders and rubbing strips to the fender piles will be installed.

2.3.4   The caisson for the breakwater

Caisson breakwater

The construction of the caisson is covered in the Bill of Quantities by item 5.2.

This assumes that the caisson will be constructed *in situ*. A cheaper solution is proposed by the prefabrication of the caissons and then towing them to their final location. This method of construction is now described in detail.

(a)   *The fabrication of the caisson*

(i)   Caisson yard

There being no drydock or caisson yards near the site, caissons shall be fabricated on the existing land as shown in figure 9. The subsoil of the fabrication yard is a fine sand layer and there will be sufficient bearing capacity to resist the weight of caisson. In order to achieve a perfectly flat and horizontal surface, and to avoid any uneven settlement, a 100 cm thickness of sand-gravel mixture will be spread and compacted to achieve a horizontal surface.

(ii) Fabrication

Scaffolding shall be provided inside and outside of the caisson by the use of scaffolding tubes and fittings.

Concrete shall be placed in three cycles, i.e., bottom slab, first pour and second pour.

The formwork shall comprise steel shutters, which will be assembled in large panels and set in position by cranes.

Concrete will be placed by a concrete pump with a moveable boom and flexible rubber hoses.

Figure 21 shows details of the fabrication of the caissons.

(b) *Transportation and setting*
(i) Transportation

In order to shift the caisson units from the fabrication yard, caissons completed, use will be made of the dredger. As shown in figure 22, the fabrication yard will be excavated by the dredger to allow the caisson to float naturally on the sea. This method requires a long distance between the first cycle yard and the second cycle yard, to avoid any serious effect on the bearing capacity of the latter yard.

Each caisson will be individually towed into position by a 1500 PS tugboat.

(ii) Setting

Anchors, wire ropes and winches shall be prepared for setting and locating the caisson in position. The correct position will be determined by survey from two directions. When the caisson is located in the correct position, it will be flooded and lowered on to the seabed. Careful observation must be continued until the caisson is sunk in its final position. Then sand will be filled inside the caisson, the sea water being expelled by pumps.

(c) *Caisson mound*

Rubble bedding stones will be placed prior to the setting of caissons and armour stones placed after the caissons have been secured in position.

The stone will be loaded on a bottom opening barge at the temporary loading jetty and towed to the site. The barge will be located in position by survey. Immediately on reaching the correct position, the stones on the barge will be deposited on the seabed. In the case of armour stones, the stones will be placed individually using the crane barge and bucket. This is to avoid destruction of the caisson mound.

Formation of the mound will be carried out by divers and underwater bulldozers, checking location, elevation and gradient of the slope.

Figure 21.   Fabrication of the caissons

a. Inside scaffolding.
b. Outside scaffolding
c. Formwork
d. Tiles
e. Spacer
f. Connection to wall
g. Sand and vinyl sheet

1. Pour 1 (soffit slab)
2. Pour 2 (lower wall)
3. 3rd concreting (upper wall)

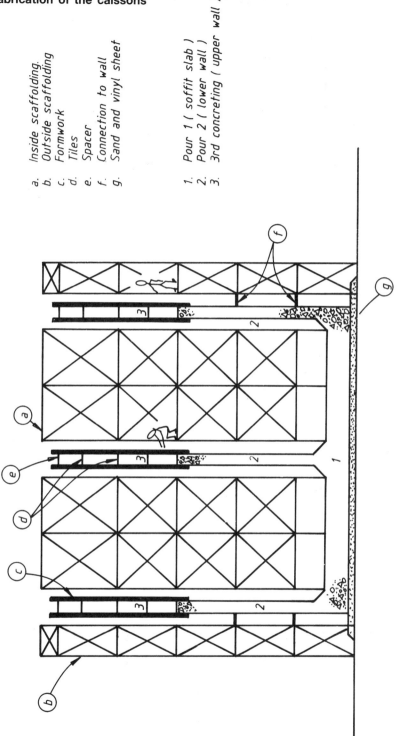

Figure 22.   Transportation of the caissons

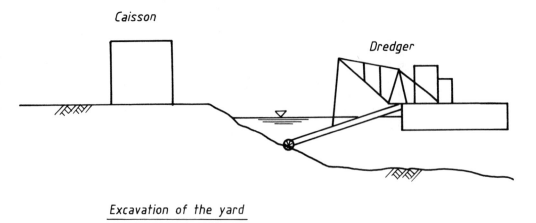

Caisson

Dredger

Excavation of the yard

Natural Shifting

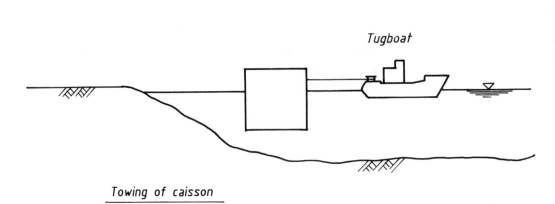

Tugboat

Towing of caisson

(d)   In situ *concrete topping to the caisson*

Formwork will be fixed by ties because no supports from outside are available. Scaffolding for this formwork will be located and secured using embedded anchor bolts in the caisson wall.

Concrete will be poured from land or from a position on the already completed caisson topping concrete slab.

2.4   The time schedule for construction

The overall time schedule for the main construction activities is shown in figure 23.

The time allowed for each of these activities is now explained.

2.4.1   Mobilization

All materials and construction plant, except for aggregate, sand and rock, will be purchased and hired in the Unified Republic and shipped to Federalstan.

The procurement of materials and plant will be started after being issued the Intent of the Award of Contract. Shipment of material and plant shall be started at or after the date of Commencement of the Works.

Estimates of the time necessary for the mobilization are as follows:

Shipment	0.5 month
Customs clearance	1.0 "
Local transportation	0.5 "
Total	2.0 months

2.4.2   Site investigation

Prior to the fabrication of prestressed concrete piles for the jetty, subsoil conditions shall be reconfirmed in order to decide pile lengths, carrying out soil investigations (boring, sampling and testing) as specified in clause 103(1) of the Specifications.

To save time, a local organization shall be employed under a subcontract and the investigation shall immediately be commenced at the date of Commencement.

One month is estimated from the start of the contract to the submission of investigation reports.

Figure 23. The time schedule for construction

Description			Quantity	Time (Calendar Month)	Notes
Mobilisation			1 LS		
Site Ivestigation			1 LS		
Site Preparation			1 LS		
Dredging			3,000.000M³		
Reclamation			100.000M³		
Slope Protection			1.000M		
Jetty	Piling		120 NOS.		
	Deck Slab		150 M		
	Fender		42 NOS.		
Break -water	Mound		62 400M³		
	Fabrication of caisson		60 NOS.		
	Setting fill topping		60 NOS.		
	Armour		37 200M³		

2.4.3 Site preparation

Site preparation shall include, but not be limited to, the following items:

(a) *Administration*
 - customers' formalities (process of import);
 - any other formalities prescribed by the law;
 - recruitment of local labour, drivers, operators, local staff and local supervisors;
 - information gathering necessary to execute the project; and
 - installation of the telephone and other services.

(b) *Construction of site buildings and facilities*

The construction of the site buildings and facilities will include:
 - accommodation for the Engineer and staff;
 - site offices;
 - batcher plant;
 - crushing, sieving, washing plant;
 - warehouses;
 - repairing factory;
 - temporary loading jetty;
 - desalination plant and water storage tanks;
 - oil and fuel storage tanks;
 - first-aid facilities, etc.

The site preparation will be started immediately following the Commencement of the Works with the aim of completion one month after the end of mobilization.

2.4.4 Dredging

(a) *Dredging period*

Details of the dredger employed in this project are as follows:

Power: Steam turbine
Pump power: 10 000 PS
Dredging capacity: 1 500 m³ /hour
Maximum distance of discharging: 8 000 m

An average hourly dredging amount of 1,000 m³ is estimated given the subsoil conditions, including contingencies such as bad weather, high winds, breakdowns, waiting time. The total time necessary to complete the dredging of 3 million m³ is calculated under the 24-hour full operation system with three shifts.

$$\text{Total time} = \frac{3\ 000\ 000}{1\ 000/\text{h.} \times 24\ \text{hrs./day}} = 125\ \text{days}$$

(b) *Preparation for dredging*

Time for towing the dredger to Federalstan and clearing customs are incorporated in the mobilization.

Preparation time (assembling ladders and cutter, laying discharge pipe, etc.) is estimated to be 0.5 month.

(c) *Dismantling and clearance*

Time for dismantling the dredging equipment and clearing the site is included in the mobilization.

2.4.5 Reclamation

The breakdown of reclamation is as follows:

Hydraulic filling: 70 000 m³
Filling in the dry: 30 000 m³

Hydraulic filling shall be achieved at the start of the dredging work. This activity is not considered critical because of a relatively small quantity involved.

Filling in the dry shall be started after a certain lapse of time from the completion of hydraulic filling, and must be completed before the commencement of pile driving for the jetty.

Maximum supply of filling material from an approved borrow pit is assumed to be 500 m³ /day. Again this supply will be subcontracted. The total time for the work is calculated as follows:

$$\text{Total time} = \frac{30\ 000\ \text{m}^3}{500\ \text{m}^3/\ \text{day}} \times \frac{7\ \text{days/week}}{6\ \text{working days}}$$
$$= 70\ \text{calendar days.}$$

2.4.6 Slope protection

Total quantity of rocks for the slope protection is 70,000 m³. Maximum supply of rocks is limited to 300 m³ per day because of output restrictions at the quarry. (The supply of this material will be subcontracted.)

$$\text{Total time} = \frac{70\ 000\ \text{m}^3}{300\ \text{m}^3/\ \text{day}} \times \frac{7\ \text{days/week}}{6\ \text{working days}}$$
$$= 272\ \text{calendar days.}$$

2.4.7 Jetty

(a) *Test piling and test loading*

This is to confirm whether or not the piling method planned, the design of piles and the estimated bearing capacity of piles match the subsoil conditions. When testing, care must be taken to check the following items:

- the hammer, cushion, and drop height of hammer;
- the visual inspection of the pile, the number of blows, penetration per blow; and
- the ultimate bearing capacities of piles.

The test includes:

- driving four piles;
- establishment and removal of staging with reaction piles;
- loading and recording data;
- removal of test piles (cut off at 1 m below the designed seabed level and dispose of).

This test shall be started upon completion of dredging in the testing area and will be completed in or about one month.

(b) *Pile driving*

The total number of piles to be driven is 120. The average time required to locate and drive one pile is estimated as follows:

– setting a pile on the leader from pontoon:	20 min.
– conducting the barge in position:	20 min.
– check of anchor, etc.:	10 min.
– driving:	30 min.
– temporary support for pile:	10 min.
– shifting:	20 min.
Total	1 hr. 50 min.

Assuming average working time per day = 9 hours;
reduction due to marine conditions and weather = 20%;

therefore, an estimated number of piles to be driven per day will be:

$$\frac{9 \text{ hrs.} \times 0.8}{1 \text{ hr. } 50 \text{ min./pc}} = 3.93 = 4 \text{ Nos./day}$$

Thus, total time to complete the piling at the jetty will be:

$$\frac{120 \text{ Nos.}}{4 \text{ Nos./day}} \times \frac{7 \text{ days/week}}{6 \text{ working days}} = 35 \text{ days}$$

(c) *Concrete deck slab*

Each panel will be constructed in two parts with a vertical construction joint between them, due to the volume of concrete, surface area to be finished and working hours. Individual working items are estimated as follows:

Falsework	2 (days)
Pile cap	1
Soffit shutter	2
Reinforcement	5
Side shutter	2
Preparation and inspection	1
Concreting	1
Curing	14
Removal/cleaning	2
Total	30 days

Total time = 1 month × 2 parts × 3 panels
= 6 months

(d) *Fender system*

The time for setting the fender system is estimated as:

Fender piles (42 Nos.)	7
Rubber fenders (42 Nos.)	4
Assembling the above with rubbing strips (14 sets)	7
Total	18 days

However, since the assembling work includes welding and painting, and production will be affected by sea conditions and inclement weather, one month is assumed.

2.4.8 Breakwater

(a) *Rubble bedding stones as the caissons' foundation*

Total net volume of the rubble stones = 62 400 m³.
A 500 m³ bottom-opening barge will be employed:

loading 500 m³	6 hrs. 30 min.
transport	30 min.
positioning	20 min.
unloading	20 min.
return	20 min.
Total	8 hrs. 00 min.

Therefore, total construction time is:

$$\frac{62\ 400\ \text{m}^3}{500\ \text{m}^3/\text{day}} \times \frac{7\ \text{days/week}}{6\ \text{working days}} = 146\ \text{days}$$

This can be commenced after dredging in that area, and formation can be done using the loading time.

(b) *Fabrication of caisson*

Materials for formwork and scaffolding with optimum gang size for each work item shall be so prepared as to be able to construct six caissons in one cycle, i.e., totalling ten cycles for 60 Nos. of caissons. In this type and size of caisson, it is estimated by past records of performance that one cycle needs approximately one month. Therefore, the total construction period will be ten months. It shall be started immediately after the site preparation work has been completed.

(c) *Transportation, setting, filling and topping*

As previously mentioned, the dredger is needed to transport the caissons from the fabrication yard. This transportation therefore cannot be started until the dredging work has been completed, i.e., month nine of the Contract. The setting schedule for the caissons will be linked to the fabrication cycle. It is assumed that ten caissons will be established per month. Filling and topping will proceed at a rate dependent upon the progress of the setting. Thus, the setting will be completed at the end of the fourteenth month. Months 15 and 16 will be spent completing the filling and topping of the last ten caissons and the other miscellaneous works.

(d) *Armour rocks*

The draft of a single caisson with no loading is estimated to be 7.12 m. If armour rocks are placed before setting caissons, the working period to set caissons is very limited due to the tidal level. This would be likely to cause a serious delay in the total construction programme. Therefore, armour rocks shall be placed after setting caissons.

The same supply of rocks and the same productivity as those for the slope protection works are expected.

The total time required is calculated as follows:

$$\frac{37\ 200\ \text{m}^3}{300\ \text{m}^3/\text{day}} \times \frac{7\ \text{days/week}}{6\ \text{working days}} = 145\ \text{days}$$

This time depends on the progress of the setting of caissons.

Thus, the construction period of armour rocks shall follow the above schedule of eight months, not 145 days.

2.5 The schedule of construction plant required

The plant required for the construction of the works has been scheduled and listed in tables 1 to 3 inclusive. This summary has been produced assuming the following:

(1) All plant required by subcontractors, such as that required for soil investigation, will be supplied by the respective subcontractors and is not included in the schedules.

(2) A schedule has been prepared of general plant items. These items of plant will be used for various job items and their costs are considered within the site on-costs not individual bill items.

(3) Both plant from the UR and local plant will be used on the project. Local plant will be required for any works within the site preparation as there will be insufficient time to transport plant to Federalstan to undertake this work.

(4) Some plant items will be purchased new in the UR and transported to Federalstan. Although these plant items must be exported on completion of the work, their costs will be written off against the project.

(5) Some plant items will be hired internally from the contractor's own plant company, others will be hired from other plant companies in the UR.

Table 1. Construction plant required (1)

Description	Items	Specification	Nos.	Period of work (month)
Mobilization and site preparation	Truck crane	20 tonne	2	1 – 3
	Derrick	100 tonne	1	1 – 3
	Trailer	20 tonne	2	1 – 3
	Bulldozer	D-5	1	1
	Back-hoe	0.8 m³	1	1 – 2
	Crawler crane	50 tonne	1	1 – 3
	Generator	175 KVA	3	1 – 3
Dredging	Dredger	10 000 PS	1	4 – 8
	Truck crane	20 tonne	1	Occasional
	Trailer	20 tonne	1	uses
Reclamation	Bulldozer	D-5	2	7
	Back-hoe	0.8 m³	1	7
	Grader	4.5 m	2	7
	Tyre roller	11 tonne	1	7
Slope protection	Crawler crane	50 tonne	1	8 – 16
	Orange bucket	1.5 m³	2	8 – 16
	Top open barge	1 000 m³	1	8 – 16
	Crane barge	50 tonne	1	8 – 16
	Tugboat	500 PS	1	8 – 16
Jetty	*Piling*			
	Piling barge		1	5 – 8
	Steam hammer	MRB 1500	1	5 – 8
	Tugboat	1500 PS	1	5 – 8
	Anchor boat	5 tonne	1	5 – 8
	Pontoon	500 tonne	2	5 – 8
	Traffic boat		2	5 – 8

Table 2. Construction plant required (2)

Description	Items	Specification	Nos.	Period of work (month)
Jetty	*Deck slab*			
	Concrete breaker	20 kg	4	9 – 12
	Pontoon	500 tonne	2	9 – 14
	Crane barge	50 tonne	1	9 – 14
	Traffic boat		2	9 – 14
	Concrete pump		2	9 – 14
	Vibrators		8	9 – 14
	Fender system			
	Crane barge	50 tonne	1	15
	Pontoon	500 tonne	1	15
	Vibration hammer	45 kW	1	15
Breakwater	*Mound*			
	Bottom open barge	1 500 m³	1	5 – 9
	Bulldozer	Underwater	1	5 – 9
	Command boat		1	5 – 9
	Generator	125 KVA	1	5 – 9
	Setting			
	Dredger	10 000 PS	1	9 – 16
	Tugboat	1 500 PS	1	9 – 16
	Winch	10 tonne	4	9 – 16
	Anchor boat	5 tonne	1	9 – 16
	Crane barge	50 tonne	1	9 – 16
	Top open barge	1 000 m³	2	9 – 16
	Pump	200 mm	4	9 – 16

Table 3. Construction plant required (3)
(These plant items required throughout the contract)

Description	Items	Specification	Nos.
General item	Batcher plant	70 m³ /hr.	1
	Crushing plant		1
	Sieving plant		1
	Washing plant		1
	Concrete transport vehicle	6 m³	6
	Concrete pump	60 m³ /hr., with	
		300 m pipes	2
	Vibrator	50 mm ∅ shaft	10
	Concrete bucket	1 m³	2
	Water tank	100 m³	2
		20 m³	2
	Desalination plant	50 m³ /day	1
	Generator	125 KVA	6
		175 KVA	2
		300 KVA	2
	Compressor	5 m³ /min.	4
	Belt conveyor	10 m	4
	Forklift	5 tonne	1
	Shovel dozer	0.6 m³	1
	Welding machine		4
	Re-bar bender	29 mm	2
	Re-bar cutter	29 mm	2
	Truck crane	20 tonne	2
	Derrick	2 tonne	2
	Trailer	20 tonne	1
	Pick-up truck	1 800 cc	4
	Minibus	20 persons	4
	Tank lorry	30 m³	2
	Gantry crane	40 tonne	1

2.6 The schedule of labour requirements

As an initial basis for sizing the workforce for the project, the following trade categories have been identified:

(a) *reinforcement benders* – skilled workers who cut and bend reinforcement according to the bending schedule;

(b) *reinforcement fixers* – skilled workers who fix reinforcement in accordance with the drawings;

(c) *carpenter* – skilled workers responsible for the fabrication, installation and removal of all formwork;

(d) *scaffolder* – skilled workers who undertake the erection and dismantling of all scaffolding and falsework together with the loading/unloading of heavy equipment;

(e) *crew* – workers on ships such as piling barge, dredger, tugboat and any other ships, except for crane barges;

(f) *concrete worker* – experienced workers for concrete placing and the preparation for it;

(g) *labour* – any unskilled workers who work under instructions of the above skilled workers;

(h) *operator* – operators of the required construction plant with acceptable qualifications and current licences;

(i) *driver* – drivers with current licences of any vehicles used for construction.

Under these definitions, the workforce for each trade are summarized in tables 4 and 5 together with the period for which they are to be engaged. The tables also indicate whether the labour will be local or expatriate labour.

Table 4. Workforce requirements (1)

Description	Category	Nos.	Period	Nationality
Mobilization	Scaffolder	5	2 – 3	UR
	Scaffolder	5	1 – 3	Local
	Operator	8	1 – 3	Local
	Driver	2	1 – 3	Local
	Labour	25	1 – 3	Local
Dredging	Crew	15	4 – 8	UR
	Operator	1	4 – 8	Local
	Driver	1	4 – 8	Local
	Labour	2	4 – 8	Local
Reclamation	Operator	6	6 – 7	Local
	Labour	6	6 – 7	Local
Slope protection	Crew	1	8 – 16	UR
	Operator	2	8 – 16	Local
	Diver	2	8 – 16	Local
	Labour	3	8 – 16	Local
Jetty				
(1) *Piling*	Crew	8	5 – 8	UR
	Operator	1	5 – 8	Local
	Labour	3	5 – 8	Local
(2) *Deck slab*	Re-bar bender	2	9 – 14	UR
	Re-bar fixer	2	9 – 14	UR
	Re-bar bender	2	9 – 14	Local
	Re-bar fixer	2	9 – 14	Local
	Scaffolder	2	9 – 14	UR
	Operator	1	9 – 14	Local
	Carpenter	4	9 – 14	UR
	Labour	15	9 – 14	Local
(3) *Fender*	Scaffolder	2	15	UR
	Labour	3	15	Local

Table 5. Workforce requirements (2)

Description	Category	Nos.	Period	Nationality
Caisson	Re-bar bender	2	4 – 13	UR
		2	4 – 13	Local
	Re-bar fixer	2	4 – 13	UR
		2	4 – 13	Local
	Scaffolder	15	4 – 13	Local
	Operator	1	4 – 13	Local
	Crew	5	9 – 16	UR
	Carpenter	10	4 – 13	UR
		20	4 – 13	Local
	Diver	2	9 – 16	UR
	Labour	10	4 – 16	Local
General	Welder	2	4 – 14	UR
		2	1 – 18	Local
	Operator	2	4 – 14	UR
		2	1 – 18	Local
	Driver	8	1 – 18	Local
	Mechanic	2	1 – 18	Local
	Electrician	2	1 – 18	Local
	Concrete worker	8	4 – 18	Local
	Labour	4	4 – 18	Local
	Watchman	6	1 – 18	Local
	Housekeeper	3	1 – 18	Local
	Office boy	2	1 – 18	Local

ESTIMATING THE COST OF CONSTRUCTION

3

3.1 Establishing costs

3.1.1 The cost of materials

It is necessary to obtain a quotation for all the main materials required for the construction work and those required to establish and maintain the site. This requires a detailed materials schedule to be prepared and the total quantities to be abstracted from the contract documents. Having established the materials required, quotations for the goods will be sought from suitable suppliers. It is necessary to consider more than the basic price for the goods.

The inquiries should include:

- the specification of the material;
- the quantity of the material;
- the likely delivery programme including the period for which supplies will be needed;
- the address of the site or bond warehouse and the port from which materials will be exported;
- the means of packing;
- the period for which the quotation remains either open for acceptance or firm;
- the date by which the quotation is to be submitted; and
- the general trading conditions.

The materials prices to be used in the project are listed in tables 6 to 8 inclusive.

Table 6 shows the materials to be purchased locally in Federalstan. The prices shown are in UR dollars to enable the material costs to be easily included within the estimate. (An exchange rate of 30 Nu equal to 1 UR dollar has been assumed.) These prices are considered to cover all transportation costs to site and any local taxes or duties.

Table 6.   Materials purchased in Federalstan

Description	Specification	Unit price (UR dollars)	Unit
Cement	Sulphate resist.	56.00	Tonne
Reinforcement	Up to 10 mm	326.00	Tonne
	Up to 20 mm	327.50	Tonne
	Up to 32 mm	331.42	Tonne
Tie wire	No. 21	480.00	Tonne
Sand	For filling	8.00	m³
Aggregate		15.00	m³
Armour rock	200–300 kg/pc	18.00	m³
Rubble stone	50–100 kg/pc	20.00	m³

Note: Sand and quarry delivered on site in the above price. Armour rock and rubble stone are free on barge.

Table 7.   Permanent materials purchased in the UR

Description	Specification	Unit price (UR dollars)	Unit
PC piles	Ø 750 * 125 mm	960.00	pc
	Ø 900 * 150 mm	1 840.00	pc
Fender pile	Ø 300 * 9 mm	550.00	pc
Admixture		1.05	m³
Filter fabric	t = 15 m	3.00	m²
Rubber fender	cell type	2 500.00	pc
Rubbing strip	Steel with anti-corrosion coating (W = 3 ton/pc	2 500.00	pc
50 t bollard	Cast iron	300.00	pc
Ladder	Galvanized	1 000.00	pc
Steel angle	Galvanized 150 * 150 * 10	300.00	Tonne
Buffer plate	With anchors and shock absorber	500.00	pc
PVC pipe	Ø 150 mm	0.70	m
Grout		600.00	m³

Note: All prices quoted are Free on Board prices.

Table 8. Temporary materials purchased in the UR

Description	Specification	Unit price (UR dollars)	Unit	Notes
Tube	Ø 48.3 mm	1.20	m	
Pipe support	4–5 m	50.00	pc	
Joint		0.60	pc	
Base plate		1.00	pc	
Coupler	Double, swivel	1.00	pc	
H-beam	305 × 305 × 10 × 15	320.00	Tonne	
Channel	300 × 90 × 12	300.00	Tonne	
Channel	305 × 102 × 12	304.00	Tonne	
Angle	100 × 100 × 10	239.00	Tonne	
Steel bracket		1 000.00	Tonne	
Steel plate	t = 9 mm	400.00	Tonne	
Scaffold board	4 000 × 250 × 24	5.00	pc	
Metal form	1 500 × 30	10.00	pc	
Curved metal		40.00	pc	
Plywood	t = 9 mm, 4' × 8'	17.94	pc	$6.23/m^2$
Timber	Processed	100.00	m^3	
Tie	Ø 9 mm, 40 cm	1.00	pc	
Chamfer	20 mm × 20 mm	0.25	m	
Oil for formwork		1.20	litre	
Curing mat	t = 10 mm	3.00	m^2	
Curing compound		2.00	kg	
PVC pipe	Ø 150 mm	2.00	m	
Oxygen		1.00	kg	
Acetylene		3.00	kg	
Office		100.00	m^2	
Accommodation		50.00	m^2	
Warehouse		3 000.00	pc	
Laboratory		50.00	m^2	

Note: All prices shown are Free on Board prices.

Tables 7 and 8 show the temporary and permanent materials to be purchased in the UR and shipped to Federalstan. The prices shown are Free on Board (FOB), prices quoted in UR dollars.

FOB prices are those costs which are quoted by the supplier and include all costs of the goods up to the point where they are received over the ship's rail at the port of departure (i.e., the cost of the goods ex-works plus all transportation costs, harbour dues, loading costs, etc.).

Having established the FOB prices the estimator must make a decision with respect to how the costs of transporting these materials to site are included within the estimate. These costs will include:

- freight charges;
- carriage charges;
- insurance cost;
- import duties;
- local agent's commission;
- customs clearance;
- port handling; and
- transportation to site.

The estimator has three possible approaches to the calculation of these additional costs:

(i) to calculate these costs for *each* material and to include these costs in the unit rate for *each* material within the estimate;

(ii) to calculate these costs for a *range* of materials so producing a percentage "uplift" on the FOB prices that may be used to calculate a cost on site for all the materials required on site; and

(iii) to estimate the total shipment costs for all the materials to be transported from the UR as a single calculation and include these costs in the site mobilization costs.

The approach selected will depend on not only the time and data available to the estimator. An element of risk is involved. If these costs are included within individual bill item rates then, should there be an *increase* in the item quantity required, then payment for these additional costs on the material element is safeguarded. However, if the cost of shipment is included within the mobilization and there should be a *decrease* in the item quantity required, then additional monies over and above the actual costs are secured.

For the purposes of this case study approach (iii) above is selected. Reference should therefore be made to the calculation of mobilization costs in Chapter 3, section 3.1.4.

3.1.2 Plant costs

The policy of the contractor is to purchase new construction plant for the project and to export this plant to Federalstan. This newly purchased plant will be supplemented, where applicable, with plant hired from the contractor's own plant company. Additional plant will also be hired in Federalstan to meet short-term plant requirements. The availability of these locally hired plant items and appropriate rates are obtained from the company's local agent in Federalstan.

Tables 9 to 11 contain the costs of all three categories of plant to be used on the project. As with the materials for the project, the transportation costs of those plant items imported to Federalstan from the UR are covered in the mobilization costs.

Table 9. Construction plant and equipment to be purchased in the UR

Description	Specification	Unit price (UR dollars)	Unit	Notes
Batcher plant	70 m³/hr	200 000.00	Set	
Crushing plant		50 000.00	Set	
Sieving plant		30 000.00	Set	
Washing plant		30 000.00	Set	
Desalination plant	50 m³/day	50 000.00	Set	
Gantry crane	40 tonne	100 000.00	Set	
Generator	300 KVA, diesel	35 000.00	pc	
Minibus	20 persons	12 000.00	Nos.	
Pick-up truck	1800 cc	7 000.00	Nos.	
Laboratory equipment	Testing devices	10 000.00	L.S.	
Water pump	8" dia.	1 500.00	Nos.	
Grouter		8 000.00	Nos.	
Concrete transport vehicle	6 m³ loading	15 000.00	Nos.	Not new
Vibrator	200 mm ∅	400.00	Nos.	
Truck crane	100 tonne	160 000.00	Nos.	
Truck	4 tonne	10 000.00	Nos.	
Fork lift	5 tonne	10 000.00	Nos.	

Table 10. Construction plant and equipment to be hired in the UR

Description	Specification	Hire rate (UR dollars per month)	Notes
Dredger	10 000 PS	200 000	
Piling barge	600 PS	60 000	Leader ±30°
Tugboat	1 500 PS	12 000	
	500 PS	6 000	
Anchor boat	5 tonne	5 000	
Command boat		1 500	
Pontoon	500 tonne	3 000	
Crane barge	50 tonne	10 000	
Bottom open barge	1 500 m³	7 000	
Top open barge	1 000 m³	4 500	
Bulldozer	Underwater	3 500	
Traffic boat	300 KVA	1 000	
	175 KVA	600	
	125 KVA	400	
Compressor	5 m³/min.	500	
Winch	50 tonne	300	
Loading jack	250 tonne	800	
Steam hammer	15 tonne	1 000	
Breaker and cutter	1.5 m³/min.	1 500	
Vibration hammer	45 kW	800	
Crane	100 tonne	10 000	
Bucket		1 000	

Table 11. Construction plant and equipment to be hired in Federalstan

Description	Specification	Hire rate (Nu per month)	Hire rate (UR dollars/month)
Truck crane	20 tonne	15 000	500
Derrick	2 tonne	6 000	200
Trailer	20 tonne	4 500	150
Bar bending equipment		2 250	75
Crawler crane	50 tonne	27 000	900
Bulldozer	D–5	36 000	1 200
Back-hoe	0.8 m³	10 500	350
Grader	4.5 m	15 000	500
Tyre roller	11 tonne	12 000	400
Shovel	0.6 m³	13 500	450
Tank lorry	30 m³	15 000	500
Generator		6 000	200
Concrete pump	60 m³/hr	8 000/day	270/day

3.1.3 Labour costs

Both "local", Federalstan labour, and UR workers are to be used for the project. Tables 4 and 5 in section 2.6 of Chapter 2 show the breakdown of the local and expatriate labour.

Information on "local" labour costs was obtained for use in the estimate by the company's agent in Federalstan and the availability of the labour was checked by the contractor's staff on the site visit. The labour costs assumed for local labour include all allowances, insurances, transportation and site accommodation costs over the 18-month duration of the project. Payment for local labour will be in the currency of Federalstan, the Nu. For the purposes of the estimate these costs are converted into UR dollars. These costs are shown in table 12.

For the expatriate labour the monthly costs calculated assume consideration of all the following elements:

- basic time;
- overtime;
- personal insurance;
- sickness pay;
- paid leave allowance; and
- end-of-project bonus.

Table 12. Labour costs: Federalstani and UR labour

Occupation	Federalstani (UR dollars/month)	UR (UR dollars/month)
Supervisor	35	2 000
Operator	117	
Driver	50	
Carpenter	100	1 500
Reinforcement bender, fixer	84	1 500
Scaffolding	84	1 500
Labour	50	
Mechanic	167	
Electrician	167	
Diver	200	3 000
Welder	150	1 500
Watchman	34	
Office boy	34	
Crew		2 000

3.1.4 Mobilization and demobilization costs

(a) *Mobilization*

Mobilization costs include the cost of transporting to the site the main plant, equipment and materials required to establish the site and undertake the main part of the works. Therefore, mobilization costs must include consideration of the following cost elements:

- transportation to port of departure (by supplier);
- storage at bonded warehouses;
- customs clearance;
- loading;
- shipment;
- unloading;
- customs clearance; and
- transportation to the site.

The costs for mobilization are generally divided into the following three categories:

(i)   transportation to port and shipment costs;
(ii) customs duties; and
(iii) inland transportation costs.

(i) *Transportation to port and shipment costs*

Suppliers will quote prices for materials and plant as Free on Board, FOB, which includes all costs up to and including loading charges at the port of departure. Shipment costs are based on a specified freight rate which includes allowances for documentation and currency adjustments. The cost of insurance must also be added.

(ii) *Customs duties*

Customs duties will normally need to be paid on the importation of all materials and construction plant. For this project the Federalstan Government require the payment of duties in each of the following categories:

- import duty;
- sales tax; and
- surcharge.

These are expressed as percentage of invoice value or a certain rate against materials quantity.

Customs duties are calculated by the following:

$$C = A \times (I + T + S)$$

where
- C : customs import duty
- I : import duty
- T : sales tax
- S : surcharge
- A : total value including carriage insurance and freight (CIF) value.

For different materials there are different tariff rates. These are currently as follows:

steel:	$I = 100\%$	$T = 12.5\%$	$S = 12.5\%$
material other than steel:	$I = 80\%$	$T = 12.5\%$	$S = 20\%$
construction plant:	$I = 20\%$	$T = 12.5\%$	$S = 20\%$

Under Federalstani law there is a drawback system on customs duties paid.

The amount of money to be paid back to the contractor under the above system is dependent upon the period during which temporary materials and construction plant are employed for the project. The refund rates are as follows:

Employed period duties paid	Refund rate to customs
Up to 12 months	80%
12–24 months	60%
24–36 months	40%
36 months or more	20%

(iii)   *Inland transportation costs*

These costs will include all the charges that have to be paid to deliver the goods to the site. These charges will include:

– customs clearance;

– port handling charges;

– local agent's commission; and

– transportation to site.

The mobilization costs for the construction plant and materials required for the project are now calculated.

(iv)   *Construction plant newly purchased in the UR*

The construction plant shown in table 13 will be purchased in the UR and transported to Federalstan. After the completion of the works, the plant will be transported back to the UR.

The estimator therefore needs to calculate the total value of the plant transported to site including carriage, insurance and freight charges for delivery to site. The carriage costs have been calculated on the basis of the "freight tonne", the equivalent to the actual weight of the material in tonnes or the volume of the material in cubic metres, whichever is the larger.

The costs of the plant shown in table 13 are all Free on Board (FOB) costs which have been quoted by the supplier and include all costs for the plant received over the ship's rail at the port of departure.

The additional charges relating to the transportation of the plant to site have been calculated as follows:

FOB value of the plant	=	931 000 dollars
Total freight tonnes for transportation	=	1 620
Freight rate	=	95.00 dollars/ freight tonne

Table 13. Construction plant purchased new in the UR and transported to Federalstan

Description	Capacity	Quantity	Weight (tonne)	Volume (m³)	Value (UR dollars)
Truck crane	100 t	1	150		160 000.00
Gantry crane	40 t	1		200	100 000.00
Batcher plant	70 m³/hr	1		500	200 000.00
Crushing plant		1		200	50 000.00
Sieving plant		1		50	30 000.00
Washing plant		1		50	30 000.00
Desalination	50 m³/day	1		150	50 000.00
Concrete car	6 m³	6	72		90 000.00
Truck	4 t	2	11		20 000.00
Pick-up truck	1800 cc	5	9		35 000.00
Minibus	20 persons	4	10		48 000.00
Generator	300 KVA	2	10		70 000.00
Forklift	5 t	1		40	10 000.00
Grouter		1		3	8 000.00
Laboratory equipment		1		10	10 000.00
Welding machine		4		3	8 000.00
Pump	200 mm dia.	8		2	12 000.00

For packing 10 per cent increased:
262 + 1 209 = 1 471
1 471 × (1 + 10%) = 1 620

931 000.00

Ocean freight charge	=	95 × 1 620
	=	153 900 dollars
Additional allowance for oil costs	=	Nil
Forwarding agent's commission	=	0.75% × 153 900
	=	1 154
Number of shipments	=	1
Document charges	=	1 × 380 = 380

Local harbour dues are included within the FOB cost.

Gross carriage and freight charges
= 153 900 + Nil + 1 154 + 380
= 155 434

The total value of the goods including carriage and freight (C and F)
= 931 000 + 155 434
= 1 086 434

Insurance cost = 0.50% × total value including carriage and freight
= 0.50% × 1 086 434
= 5 432

Total value including carriage, insurance and freight (CIF)

= 1 086 434 + 5 432

= 1 091 866

Import duty

The import duty on construction plant in Federalstan is calculated as follows:

$$C = A \times (I + T + S)$$

where
C : customs import duty
I : import duty
T : sales tax
S : surcharge
A : total value including carriage insurance and freight (CIF) value.

For plant: I = 20%   T = 12.5%   S = 20%

Therefore, the customs import duty:

= 1 091 866 × (0.20 + 0.125 + 0.20)

= 1 091 866 × 0.525

= 573 229

As the construction plant is only to be held in the country for a temporary period of no more than 24 months a refund of 60 per cent on the tax paid is applicable. Therefore, the customs import duty due on the plant:

= 573 229 × 0.60

= 343 937

Customs clearance	= 1 500 Nu per consignment plus 75 Nu per freight tonne: (i.e., 50 dollars per consignment plus 2.5 dollars per freight tonne)
	= (1 620 × 2.5) + 50
	= 4 100
Port handling in Federalstan	= 93.75 Nu (i.e., 3 125 dollars) per freight
	= 3 125 × 1 620
	= 5 062
Transportation to site	= 103.14 Nu (i.e., 3 438 dollars) per freight tonne

	=	1 620 × 3 438
	=	5 570
Local agent's commission	=	75 Nu (i.e., 2.50 dollars) per freight tonne
	=	1 620 × 2.50
	=	4 050
Total local charges	=	343 937 + 4 100 + 5 062 + 5 570 + 4 050
	=	362 729
The total cost of the plant including carriage, insurance and freight delivered to site	=	1 091 866 + 362 729+ 343 937
	=	1 798 532 dollars
Therefore, the mobilization cost of the plant	=	1 798 532 − FOB value of the plant
	=	1 798 532 − 931 000
	=	867 532 dollars

(v) *The contractor's internal plant*

These plant items already belonging to the contractor's organization will need to be transported to the site and temporarily imported to Federalstan for the duration of the contract. (See table 10 for a list of these plant items.)

The contractor estimates that the total value of these plant items at the start of the project will be 3 million dollars.

Because of the nature of these marine plant items it is estimated that the plant and equipment may be transported to the site with the assistance of two tugboats. A quotation of 92,000 dollars is received for this transportation.

An insurance quotation of 0.5 per cent of the carriage and freight value is received. Therefore:

The insurance cost	=	(3 000 000 + 92 000) × 0.5%
	=	15 460
The total CIF value	=	3 092 000 + 15 460
	=	3 107 460
The customs import duty, calculated as previously	=	3 107 460 × 0.525 × 60%
	=	978 849
Customs clearance charges are estimated as	=	5 000 dollars

There will be no transportation to site costs.

The total cost of the contractor's internal plant delivered to site is therefore

= 92 000 + 978 849 + 5 000

= 1 075 489

(vi) *Materials*

The principal materials required for the contract are shown in table 14.

Table 14. **Materials to be purchased in the UR and transported to Federalstan**

Description	Quantity	Weight (tonne)	Volume (m³)	Value (UR dollars)
Pc pile ⌀ 750	98 Nos.	1 152	1 080	94 080
Pc pile ⌀ 900	26 Nos.	549	513	47 867
Steel pipe pile ⌀ 300	42 Nos.	49	66	23 100
Forms				
(a) Metal form	2 000 Nos.	284	450	20 000
(b) Plywood	1 000 Nos.	28	36	17 940
(c) Miscellaneous	1 L.S.	2		
Falsework				
(a) Tube	2 000 m	9	5	2 400
(b) H-beam	5 320 m	500	480	160 025
(c) Pipe support	500 Nos.	7.5		25 000
(d) Steel bracket	32 Nos.	2		600
Scaffolding				
(a) Tube	15 000 m	66	38	18 000
(b) Joint	2 000 Nos.	1		1 200
(c) Base plate	1 500 Nos.	2		1 500
(d) Coupler	5 000 Nos.	6		5 000
Others				
Admixture		21		30
Fender	42 Nos.		42	105 000
Rubbing strip	14 Nos.	28		35 000
Processed timber	300 Nos.		150	15 000
Bollard	12 Nos.	2	1 000	69 000
		Total = 4 498.5 freight		640 742

These materials are estimated to total 4,498 freight tonnes including a 10 per cent allowance for packing.

The FOB value of these materials is 640,742 dollars.

The total carriage, insurance and freight costs of these materials is calculated in a similar manner to that of the purchased plant. This is calculated to be 110,719 dollars.

The customs import duty on the material will vary according to the type of item, steel having a different import duty to that of other materials.

From table 14, it is estimated that approximately 70 per cent of the material's value is steelwork. Therefore, the customs duty for this material:

= 0.70 × (640 742 + 110 719) × (1.0 + 0.125 + 0.125)
= 0.70 × 751 461 × 1.25
= 657 528.

For the other materials, the customs duty:

= 0.30 × (640 742 + 110 719) × (0.80 + 0.125 + 0.20)
= 0.30 × 751 461 × 1.125
= 253 618.

From table 7 it is apparent that the prestressed concrete piles and the steel fender piles are the only permanent materials. The remainder of the material is re-exportable and therefore liable to a 60 per cent drawback on the taxes paid. This is estimated to produce a saving of 391,492 dollars.

The total cost of the materials listed in table 14 as delivered to the site is therefore:

= 640 742 + 110 719 + 657 528 + 253 618 − 391 492
= 1 271 115 dollars.

The cost of mobilization of these materials is therefore:

= 1 271 115 − 640 742
= 630 373 dollars.

The total cost of mobilization.

The total cost of mobilization is then calculated as follows:

Newly purchased construction plant	=	867 532
Contractor's internal plant	=	1 075 489
Materials	=	630 373
Total	=	2 573 394

A reduction on this cost is available as, under the contract, the Government of Federalstan undertake to reimburse all duties on construction plant subject to the supply of a bank guarantee covering the sum involved.

Therefore, the mobilization costs:

= 2 573 394 − 343 937 − 978 849
= 1 250 608 dollars.

The following factors also apply to the mobilization:

(i) all plant items are shown to be mobilized at the start of the contract; and

(ii) the customs duty is repayable on the provision of a bank guarantee redeemable if the plant is not exported at the end of the project. This reimbursement should be treated as a local cost saving.

(b) *Demobilization*

The estimator must calculate not only the cost of transporting all the plant and equipment to site but also the cost of transporting the plant and temporary materials back to the UR. These costs are calculated in a similar manner to mobilization costs but based on the reduced value of the plant and equipment.

For the purposes of this case study it is assumed that the demobilization costs for the plant and equipment totals 500,000 dollars.

Total mobilization and demobilization costs:

= 1 250 608 + 500 000
= 1 750 608 dollars.

3.2 The calculation of the direct cost of the work

3.2.1 Unit rate calculations

In order to calculate the direct cost of the works a number of unit rates are calculated. Examples of these calculations are now given.

(a) *Concrete (mixing and transportation)*

The cost includes:

− purchase of the sand, cement and aggregate;
− transportation of materials to site;
− handling and storage;
− mixing; and
− transportation of concrete on site.

The total quantity of concrete required to be mixed and placed is calculated to be 30,000 m³. A wastage allowance of 5 per cent on sand, coarse aggregate and cement is assumed. All materials for the concrete are to be purchased locally. Local labour will be used to mix and transport the concrete on site. The plant for the project will be purchased new in the UR and temporarily imported to Federalstan.

Material cost

The material cost per m³
of concrete is calculated as:

Cement 360 kg/m³ @ 56.00/ tonne	=	20.160
Sand 636 kg/m³ @ 8.00/m³ /2.62 = 0.636 × 8.00/2.62	=	1.942
Coarse aggregate = 1 195 kg/m³ @ 15.00/m³ = 1 195 × 15.00/2.65	=	6.764
Admixture 1.00 kg/m³ @ 1.05/kg	=	1.050
	=	29.916
Allowance for wastage (5%)	=	1.496
Allowance for water (mixing, washing, curing etc.)	=	0.700
Material cost of concrete/m³	=	32.112

Material cost subtotal
= 30 000 × 32.112
= 963 360 dollars

Plant cost

Batcher plant	=	200 000
Crushing plant	=	50 000
Sieving plant	=	30 000
Washing plant	=	30 000
Transportation vehicle 15 000 × 6 Nos.	=	90 000
Belt conveyor 1 000 × 6 Nos.	=	6 000
Generator 1 000 × 18 months	=	18 000
Fork lift	=	10 000
Plant subtotal:	=	434 000

Labour

Operator:	117 × 15 months × 1	=	1 755
Driver:	50 × 15 months × 6	=	4 500
Labour:	50 × 15 months × 8	=	6 000
Labour subtotal:		=	12 255

Total material, plant and labour
cost
=   963 360 + 434 000 + 12 255
=   1 409 615 dollars

Unit rate of concrete   =   1 409 615

                                                   30 000 m^3
                                           =   46.98
                                           =   47.00/m^3

The rate is then divided into the "local" and "offshore" elements:

	Local element	Offshore element
Labour	0.4/m^3	Nil
Plant	Nil	14.48/m^3
Material	32.11/m^3	Nil

(b)   *Pile driving*

This work relates to the driving of the prestressed concrete piles and steel piles to the jetty. The plant used for this work is the plant hired internally from the contractor's own plant company. The labour required for the work will consist of both local and expatriate labour.

The work involved in the piling operation is described in Chapter 2, section 2.4.7. One month has been allowed for the test piling and one month for the main piling operation. The whole operation will take place between months five and eight and hence the plant will be utilized on the contract for four months. The plant required is included in table 10. These items of plant are costed as follows:

Plant costs

Piling barge	60 000 × 4 months × 1 No.	= 240 000
Crane barge	10 000 × 4 months × 1 No.	= 40 000
Pontoon 500 t	3 000 × 4 months × 2 No.	= 24 000
Tugboat 500 PS	12 000 × 4 months × 1 No.	= 48 000
Anchor boat	5 000 × 4 months × 1 No.	= 20 000
Traffic boat	1 000 × 4 months × 2 No.	= 8 000
Steam hammer	1 000 × 4 months × 1 No.	= 4 000
Generator	1 000 × 4 months × 1 No.	= 4 000
Fuel and maintenance 10% × 388 000		= 38 800

Subtotal   = 426 800
                                                                   dollars

Labour costs

Crew	$2\,000 \times 4$ months $\times 8$	=	64 000
Operator	117×4 months $\times 1$	=	468
Labourer	50×4 months $\times 3$	=	600
Subtotal		=	65 068

Total labour and plant cost for the piling:

= 426 800 + 65 068

= 491 868 dollars

Three sizes of pile will be driven: 300 mm diameter steel; 750 mm diameter concrete and 900 mm diameter concrete. Assuming that the driving factors for the piles are as follows:

300 mm diameter steel	=	1.0
750 mm diameter concrete	=	2.0
900 mm diameter concrete	=	2.4

The cost of driving is apportioned as follows:

300 mm diameter steel pile

$$= \frac{1.0 \times 42 \text{ No.} \times 491\,868}{(1.0 \times 42 + 2.0 \times 96 + 2.4 \times 24)}$$

$$= \frac{42 \times 491\,868}{42 + 192 + 57.6}$$

$$= \frac{42 \times 491\,868}{291.6}$$

= 70 845

= 70 845/42

= 1 687 dollars per pile

750 mm diameter concrete piles

$$= \frac{2.0 \times 96 \times 491\,868}{(1.0 \times 42 + 2.0 \times 96 + 2.4 \times 24)}$$

$$= \frac{192 \times 491\,868}{291.6}$$

= 323 863

= 323 863/96

= 3 374 dollars per pile

900 mm diameter concrete piles

$$= \frac{2.4 \times 24 \times 491\,868}{(1.0 \times 42 + 2.0 \times 96 + 2.4 \times 24)}$$

$$= \frac{57.6 \times 491\,868}{291.6}$$

$$= 97\ 160$$
$$= 97\ 160/24$$
$$= 4\ 048 \text{ dollars per pile}$$

Check total cost

$$= 1\ 687 \times 42 + 3\ 374 \times 96 + 4\ 048 \times 24$$
$$= 70\ 854 + 323\ 904 + 97\ 152$$
$$= 491\ 910 \text{ (this is acceptable assuming rounding errors)}$$

(c) *Reinforcement*

The rate for reinforcement includes:
- material purchase costs;
- transportation costs;
- bending and fixing costs;
- wastage allowance; and
- additional material (spacers and tie wire) allowances.

All reinforcement will be purchased in Federalstan. Local steel fixers will be used. The following allowances will be made:

(i) A wastage allowance of 5 per cent.

(ii) The consumption of tie wire (# 21) shall be as follows:

up to 10 mm dia.	20 kg/tonne
up to 20 mm dia.	15 kg/tonne
up to 32 mm dia.	5 kg/tonne

(iii) Bending rate:

up to 10 mm dia.	30 kg/tonne
up to 20 mm dia.	15 kg/tonne
up to 32 mm dia.	10 kg/tonne

(iv) Fixing rate:

up to 10 mm dia.	0.4 tonne/day/person
up to 20 mm dia.	0.5 tonne/day/person
up to 32 mm dia.	0.6 tonne/day/person

The total reinforcement to be fixed in the works is calculated as follows:

Jetty	270 tonne
Caisson	1 796 tonne
Topping	334 tonne
Total	2 400 tonne

This reinforcement will be fixed over a 13-month period.

Plant costs

As general construction site equipment, it is assumed that the bar bending and transportation equipment will be required for the duration of the project (18 months).

The plant rate for transportation of reinforcement on site:

=	cost of truck crane plus trailer	
=	500 + 150 = 650 dollars/month	

Bar bending machinery	=	150 dollars/month
Total plant costs	=	800 dollars/month × 18
	=	14 400 dollars
Plant costs per tonne	=	14 400/2 400
	=	6 dollars/tonne

Labour costs

Gangs of local steel fixers will be employed at a rate of 84 dollars per month. (The labour will work 30 days per month totalling 200 working hours per month.)

Labour costs for transportation

Plant = 117 + 50	=	167 dollars/month
Total labour transportation costs	=	167 × 18
	=	3 006 dollars
Labour transportation costs/tonne	=	3 006/2 400
	=	allow 2 dollars/tonne

Purchase price of the reinforcement is as follows:

up to 10 mm dia.	362.00/tonne
up to 20 mm dia.	327.50/tonne
up to 32 mm dia.	331.42/tonne

Calculation of total cost rates

From the above data the total rate for each diameter of reinforcement is calculated as follows:

(a) Up to 10 mm:

reinforcement 362.00/tonne × (1 + 5%)	=	380
tie wire 0.02 tonne/tonne × 480	=	10
spacer 1.5% × reinforcement	=	6
bending 84/month × 30 hrs./200 hrs.	=	13
fixing 84/month × 1/0.4 × 1/30	=	7

transportation labour cost		=	2
transportation and bending equipment		=	6
Cost, dollars/tonne		=	424

(b) Up to 20 mm:

reinforcement 327.50 × (1 + 5%)	=	344
tie wire 0.015 tonne/tonne × 480	=	7
spacer 1.5% × reinforcement cost	=	5
bending 84/month × 15 hrs/200 hrs	=	7
fixing 84/month × 1/0.5 × 1/30	=	6
transportation labour cost	=	2
transportation and bending equipment	=	6
Cost, dollars/tonne	=	377

(c) Up to 32 mm:

reinforcement 331.42 × (1 + 5%)	=	348
tie wire 0.005 tonne/tonne × 480	=	3
spacer 1.5% × reinforcement cost	=	5
bending 84/month × 10 hrs./200 hrs.	=	5
fixing 84/month × 1/0.6 × 1/30	=	5
transportation labour cost	=	2
transportation and bending equipment	=	6
Cost, dollars/tonne	=	374

These rates are then split into the "local" and "offshore" element of cost. For example, for the rebar up to 10 mm.

	Local element	Offshore element
Labour	20	Nil
Plant	8	Nil
Material	396	Nil
Total	424	Nil

3.2.2 The calculation of the direct cost of individual bill items

This section contains examples of the calculations undertaken to price the items in the Bill of Quantities and hence establish the direct cost of the works. The labour, plant and materials rates used are those found in tables 9 to 14. Use is made of the unit rates calculated in section 1 of this chapter. The Bill of Quantities document to be priced is included in Chapter 1, section 1.4.

All calculations are made in dollars. The estimator also calculates for each item the "local" element of the works to identify which costs must be made in local currency.

(a) *Bill item 1.1:   Mobilization and demobilization*

This item is to be priced as a lump sum.

Calculations for mobilization costs are shown in 3.1.4.

The cost stated here must include both mobilization and demobilization costs.

A total sum of 1,750,608 dollars is included for this item.

(b) *Bill item 1.4:   Provision of facilities for the engineer*

This includes the following bill items:

1.4.1	Offices
1.4.2	New vehicles
1.4.3	New generators
1.4.4	Storage tanks
1.4.5	Detached houses for accommodation

The site offices will be prefabricated in the UR and erected on site. Cost of material = 100 dollars/m^2.

It is assumed that the erection of each office will take five men 40 hours. Local labour will be used at a cost of 50 dollars/month.

$$\text{Total erection cost} \quad = \quad \frac{5 \times 40}{200 \times 50} = 50 \text{ dollars}$$

An allowance of 500 dollars is entered to allow for erection costs and contingencies.

The unit cost of the new vehicles is	7 000 dollars
The unit cost of the generators is	35 000 dollars
The unit cost of the set of storage tanks is	20 000 dollars

An allowance of 20 per cent on these three items is included to cover maintenance costs. (This includes both labour and parts.)

The estimator is advised by the company's local agent in Federalstan that, using local labour and materials, a suitable detached house for a member of the engineer's staff could be built for 10,000 dollars. A rate of 12,500 dollars is entered to cover all maintenance and repair costs. The cost of these houses is a local cost element. It is decided to subcontract this element of the work.

(c) *Bill item 2.1:* *Site clearance*

    2.1.1   General site clearance   L.S.
    2.1.2   Disposal area   L.S.

These two items include the levelling of 90,000 m² of ground, the erection of 1,000 m of fencing and the construction of a temporary access road 500 m x 10 m. The estimator assumes the levelling will take two weeks, the fencing ten days and the road one week to construct. The calculation of the cost of these items is shown below. All labour, plant and materials are assumed to be obtained locally. A 5 per cent allowance on the plant costs is added to cover fuel costs.

Item	Description	Unit rate	Unit	Quantity	Time	Amount
Levelling – 90 000 m²						
Plant:	bulldozer	4 800	Month	1	0.5	2 400
	grader	4 000	Month	1	0.5	2 000
	fuel	4 400	%	5	—	220
Labour:	operator	117	Month	2	0.5	117
	labour	50	Month	2	0.5	50
					Total	4 787
Fencing – 1 000 m						
Plant:	back hoe	2 800	Month	1	0.33	924
	fuel		%	5	—	140
Material:	fencing wire	6.50 kg		650	—	4 225
	timber stakes	1	Nr	210	—	210
	timber gate	100	Nr	4	—	400
Labour:	operator	117	Month	1	0.33	39
	labour	50	Month	10	0.33	165
					Total	6 103
Temporary road – 500 m × 10 m						
Plant:	bulldozer	4 800	Month	1	0.25	1 200
	grader	4 000	Month	1	0.25	1 000
	fuel	4 200	%	5	—	210
Labour:	operator	117	Month	2	0.25	58
	labour	50	Month	10	0.25	125
					Total	2 593

The total cost of item 2.1.1 = 4 787 + 6 103 + 2 593 = 13 483 dollars

All this cost is a "local" cost.

(d) *Bill item 2.1.2:* *Disposal area*

This is a lump-sum item. The actual area to be cleared is 2,500 m².

The disposal area will be located close to the temporary construction road and the work performed simultaneously. The same unit rate is therefore assumed as for the temporary road, i.e.:

$$\frac{2\ 593}{10 \times 500} = 0.519 \text{ dollars/m}^2$$

The lump sum for this item therefore = 2 500 × 0.519 = 1 297 dollars.

(e)   *Bill item 3.1.1:   Dredging and disposal of material*

From the programme for the works, dredging, reclamation and slope protection extends over a 13-month period from month four to month 16. Assume the dredging will be completed over a six-month period. The total quantity to be dredged and disposed of = 3 million m³.

Dredging

Plant costs
Dredger at 200 000 dollars/month for 6 months = 1 200 000

Labour costs
Crew to dredger (expatriate)
3 men for 6 months at 2 000/month = 36 000

Additional local labour
1 operator at 117 dollars/month for 6 months = 702
+ 4 labourers at 50 dollars/month for 6 months = 1 200

Fuel (purchased locally)
Assume fuel cost = 5% of dredger cost
= 5% × 1 200 000 = 60 000

Total dredging cost = 1 200 000 + 36 000 + 702 + 1 200 + 60 000 = 1 297 902 dollars

Disposal

Plant costs
D5 bulldozer at 1 200/month for 6 months = 7 200

Labour costs
Operator (local) 117/month for 6 months = 702

Fuel at 5% of bulldozer cost = 5% × 7 200 = 360

Total cost of disposal = 7 200 + 702 + 360 = 8 262 dollars

Cost of dredging and disposal = 1 297 902 + 8 262 = 1 306 164 dollars

Cost of dredging and disposal per m^3
of material = 1 306 164/3 000 000 = 0.435 dollars/m^3

Total local cost element
= 702 + 1 200 + 60 000 + 7 200 + 702 + 360
= 70 164 dollars

Local cost element per m^3
of material
= 70 164/3 000 000 = 0.02 dollars/m^3

(f) *Bill item 3.3: Slope protection*

This Bill item consists of two sub-items:

3.3.1 Supply, place and shape armour   70 000 m^3
3.3.2 Supply and place filter fabric   38 500 m^2

This work is programmed to take place over a nine-month period. The marine plant required for the operation is hired internally from the company's plant division. A crawler crane will be hired locally to load the barge with the armour rock. The armour rock will be supplied under subcontract by a local supplier at a rate of 8 dollars per m^3. An additional 5 per cent is allowed for wastage. The placing of the material will require five divers for the nine-month period. It is assumed that one expatriate diver and four local divers will be employed. The total cost of these operations are calculated as follows:

Item	Description	Unit rate	Unit	Quantity	Time	Amount
3.3.1	**Armour rock**					
Material: (sub-contract)	Armour rock	8	m^3	73 500	—	588 000
Plant:	crane barge	10 000	No.	1	9	90 000
	crawler crane	27 000	No.	1	9	243 000
	open-top barge	4 500	No.	1	9	40 500
	tugboat	6 000	No.	1	9	54 000
	bucket	1 000	No.	1	9	9 000
	fuel	436 500	%	6		26 160
Labour:	crew	2 000	No.	2	9	36 000
	labour	50	No.	8	9	3 600
	operator	117	No.	1	9	1 053
					Total	1 091 313
3.3.2	**Filter fabric**					
Material:	Fabric sheeting	3	m^2	38 500		115 500
Labour:	diver (UR)	2 000	No.	1	9	18 000
	divers (local)	200	No.	4	9	7 200
					Total	140 700

The unit rate for the supply and placing of the armour rock:

= 1 091 313/70 000
= 15.59 dollars/m³.

The unit rate for the supply and placing of the fabric sheeting:

= 140 700/38 500
= 3 654 dollars/m³.

(g) *Bill item 4.1: Piling*

This major item of the works consists of the following sub-items:

4.1.1	Supply of prestressed concrete pile 750 mm dia.	96 No.
4.1.2	Supply of prestressed concrete piles 900 mm dia.	24 No.
4.1.3	Supply of steel pipe piles 300 mm dia.	42 No.
4.1.4	Handle and drive piles 750 mm dia.	96 No.
4.1.5	Handle and drive piles 900 mm dia.	24 No.
4.1.6	Handle and drive piles 300 mm dia.	42 No.
4.1.7	Test pile 750 mm dia.	2 No.
4.1.8	Test pile 800 mm dia.	2 No.
4.1.9	Static load test	4 No.

The piling materials will be purchased in the UR and shipped to site. All handling costs on site have been included within the site organization costs.

The labour and plant rates for the above items have already been calculated as unit rates in section 3.2.1.

These materials, labour and plant costs are summarized below.

(All costs are unit costs per pile.)

	300 mm dia.	*750 mm dia.*	*900 mm dia.*
Material	550	960	1 840
Labour	223	446	535
Plant	1 467	2 928	3 513
	2 240	4 334	5 888

The test loading of the piles will utilize temporary materials and equipment transported from the UR.

The cost of these materials	=	15 478 dollars
The cost of the additional labour	=	800 dollars
The cost of the additional plant	=	1 800 dollars
This gives a total testing cost	=	15 478 + 800 + 1 800
	=	18 078
The unit cost per pile for testing	=	18 078/4
	=	4 520 dollars/pile

The total cost of the piling
operation is calculated to be   =   684 380 dollars

(h)   Item 4.2.1:   *Chipping and disposal of pile tops — 750 mm dia. piles*
      and
      Item 4.2.2:   *Chipping and disposal of pile tops — 900 mm dia. piles.*

This work will take place over a one-month period.

Plant costs

Crane barge	at	10 000/month	=	10 000
Pontoon	at	3 000/month	=	3 000
Trailer	at	150/month	=	150
Truck crane	at	500/month	=	500
Breaker and cutter	at	1 500/month	=	1 500
		Total	=	15 150

Fuel at 5% of plant costs
= 5% × 15 150   =   757

		Total	=	15 907

Labour costs

Local labour costs

Operators	2	at 117/month	=	234
Labourers	10	at 50/month	=	500
Drivers	2	at 50/month	=	100
Scaffolders	4	at 84/month	=	336
		Total	=	1 170

Material costs

Scaffolding (total cost of material) = 500

Total cost of chipping and disposal of the pile tops
= 15 907 + 1 170 + 500
= 17 577 dollars

A total number of 96 + 24 = 120 piles will have to be trimmed.
Assume the time per pile is proportional to the surface area $\pi\, d^2\, /4$
where d = diameter. Surface areas:

750 mm dia. pile = $\pi \times 750^2\, /4$ = 441 843 mm^2
900 mm dia. pile = $\pi \times 900^2\, /4$ = 636 255 mm^2

Using the surface area of a 750 mm dia. pile as 1.0 gives a ratio for the 900 mm dia. pile:

= 636 255/441 843 = 1.44:1

Total nominal pile area	=	96 × 1.0 + 24 × 1.44
	=	96 + 34.56
	=	130.56

| Total cost of cutting and disposal | = | 17 577 dollars |

Total cost of cutting and disposal 750 mm dia. piles
| | = | 96 / 130.56 × 17 577 |
| | = | 12 924 dollars |

| Cost per 750 mm dia. piles | = | 12 924/96 |
| | = | 134.63 dollars per pile |

Total cost of cutting and disposal 900 mm dia. piles
| | = | 34.56/130.56 × 17 577 |
| | = | 4 653 dollars |

| Cost per 900 mm dia. piles | = | 4 653/24 |
| | = | 193.875 dollars per pile |

| Rate for 750 mm dia. piles | = | 96/130.56 × 17 577/96 |
| | = | 134.63 dollars/pile |

(i) *Bill item 4.2.9: Reinforcement*

This item consists of three sub-items:

(a) Up to 10 mm diameter
(b) Up to 20 mm diameter
(c) Up to 32 mm diameter

These items are priced directly using the unit rates calculated in 3.1.

Up to 10 mm diameter	=	424.00 per tonne
Up to 20 mm diameter	=	377.00 per tonne
Up to 32 mm diameter	=	374.00 per tonne

These rates may be subdivided as follows:

	10 mm dia.	20 mm dia.	32 mm dia.
Labour	22.00	15.00	12.00
Plant	6.00	6.00	6.00
Material	396.00	356.00	356.00
Total	424.00	377.00	374.00

3.3 The direct cost summary

The direct cost of completing the works is reached by calculating the cost of each bill item and then totalling the cost of all the bill items.

In addition to this calculation the estimator will also analyse the direct costs to identify:

- the division of the bill item costs into labour, plant, material and subcontractor elements;
- the division of the bill into "local" cost element and "offshore" cost elements;
- a reconciliation between labour and plant requirements as entered in the bill and as identified by the project programme.

An example of each of these types of analyses is now provided.

(a) *The division of the bill item rates and their analysis*

Figure 24. An example of the analysis of Bill No. 3

Item No.	Description	Labour	Plant	Materials	Subcontractor	Unit	Quantity	Schedule rate	Amount (Nu)
	Earthworks								Bill No. 3
3.1	*Dredging*								
3.1.1	Dredging as specified on the Drawings, including disposal of excavated soils	0·013	0·422	Nil	0·435	m³	3 000 000	Nil	
3.1.2	Echo-sounding after the dredging work	Nil	Nil	Nil	20 000	L.S.		20 000 [1]	
3.2	*Reclamation*								
3.2.1	Fill by hydraulic means	0·009	0·297	Nil	0·306	m³	70 000	Nil	
3.2.2	Fill in the dry	0·294	9·498	Nil	9·792	m³	30 000	Nil	
3.2.3	Levelling and topping by sand-gravel mixture	0·038	1·237	Nil	1·275	m²	52 500	Nil	
3.3	*Slope protection*								
3.3.1	Supply, place and shape armour rocks 200–300 kg	0·581	6·609	8·400	15·590	m³	70 000	Nil	
3.3.2	Supply, place in position filter fabric between rocks and the ground	0·622	0·109	2·923	3·654	m²	35 000	Nil	
									Bill Total

Figure 24 shows an analysis of bill items from Bill No. 3. Each item is analysed to show not only the total rate for the item but also the subdivision of this rate. This enables the estimator to understand how the cost of each bill item is divided.

(b) *The division of the bill into "local" and "offshore" cost elements*

Figure 25 shows an analysis of the bill items from Bill No. 3 with the monies within the bill subdivided into the "local" cost element (i.e., that money which will be paid in local currency) and the "offshore" cost element. This enables the estimator to identify the company's exposure to risk with respect to local currency payments.

Figure 25. Local and offshore cost elements   Bill No. 3

Item	Description	Total amount	Local element	Offshore element
3.1.1	Dredging	1 305 000	60 000	1 245 000
3.1.2	Echo sounding	20 000	Nil	20 000
3.2.1	Hydraulic fill	21 420	1 000	20 420
3.2.2	Fill in the dry	293 760	293 760	Nil
3.2.3	Levelling and topping	66 937	66 937	Nil
3.3.1	Armour rocks	1 091 300	89 701	1 001 612
3.3.2	Filter fabric	140 700	7 200	133 207
	Total	2 939 117	518 598	2 420 519

(c) *The plant reconciliation*

The planner and estimator have agreed between them the main plant items to be used on the project. The construction programme and plant schedules show the duration that these plant items will be on site. It is essential to check that, within the estimate, the estimator has included all the costs relating to all the plant items identified. An example of this reconciliation for a single plant item, the piling barge, is shown below.

From the construction programme, the piling barge is hired internally from the company's plant division and travels to site at the start of the project. Piling commences in month five and is completed by month eight. From month nine onwards the piling barge is used to transport the caissons to their final location. This work is completed in month 16. A minimum period of 14 days is required for the mobilization and demobilization of the barge. (For expediency a period of one month is included for mobilization and one month for demobilization.) Assuming a two-month period for travelling to site and customs clearance, the minimum time that this plant item will need to be hired is:

$= 2 + 1 + 4 + 8 + 1$
$= 16$ months.

At a hire rate of 200 000 dollars/month the hire cost
$= 16 \times 200\ 000 = 3\ 200\ 000$ dollars.

An analysis of the Bill items:

3.1.1	Dredging	(Build-up included in the text);
3.1.2	Echo sounding	(Build-up not included in the text);
5.2.1	Concrete to caisson	(Build-up not included in the text);

shows that the total number of months' hire of the barge included within these bill items is 14. This represents a shortfall of two months $\times$ 200,000/month = 400,000 dollars. This is the time the plant is on hire whilst being transported to site, and a note is made by the estimator to add this cost to both the mobilization *and* the demobilization costs. (See section 4.1(i), Adjustments to the construction cost estimate.)

This example of a plant reconciliation calculation shows how the costs as identified by the planner are reconciled with those as identified by the estimator. This will be undertaken for all the plant and key labour items by consideration of the sums of money involved as identified in the direct cost estimate. A single cost adjustment to the direct cost total will then be made or the monies distributed across several bill items.

The Direct Cost Summary and all the summary reports for the estimate are then collated and brought to the tender adjudication meeting where the decision of the tender price will be agreed.

Figures 26–33 inclusive show the direct cost of the works together with the direct cost of each bill item.

Figure 26. The direct cost of the project (1 of 7)

1 General item

Item No.	Description	Unit	Quantity	Schedule rate	Amount
1.1	Mobilization and demobilization of equipment to the site	L.S.	1	1 750 608	1 750 608
1.2	Topographic survey				
1.2.1	Survey on land	L.S.	1	10 000	10 000
1.2.2	Sounding offshore	L.S.	1	25 000	25 000
1.3	Soil investigation				
1.3.1	Onshore boring, including samples and field test	m	50	150	7 500
1.3.2	Offshore boring, including temporary staging	m	100	150	15 000
1.3.3	Laboratory test on the soil samples, including preparation of reports	L.S.	1	10 000	10 000
1.4	Provide and maintain the following for exclusive use of the engineers				
1.4.1	Offices as specified in clause 102(4)	m²	300	102	30 600
1.4.2	New vehicles for their transportation	Nos.	5	8 400	42 000
1.4.3	New generators (300 KVA)	Nos.	2	42 000	84 000
1.4.4	Potable water, oil and their storage tanks	L.S.	1	24 000	24 000
1.4.5	Detached houses for their accommodation	Nos.	5	12 500	62 500
				Bill total	2 061 208

Figure 27. The direct cost of the project (2 of 7)

2 Site preparation   Bill No. 2

Item No.	Description	Unit	Quantity	Schedule rate	Amount
2.1	*Site clearance*				
2.1.1	General site clearance	L.S.	1	13 483	13 483
2.1.2	Disposal area	L.S.	1	1 297	1 297
2.2	*Preparation of site*				
2.2.1	Site office with gate and fencing	L.S.	1	40 000	40 000
2.2.2	Site laboratory for tests of soil, cement, aggregate, concrete, water, etc., with equipment and staffing	L.S.	1	43 600	43 600
2.2.3	Warehouses for all materials used for the permanent works	L.S.	1	24 508	24 508
2.2.4	Desalination plant for potable water for living and construction purposes	L.S.	1	72 360	72 360
2.2.5	Temporary loading jetty and gantry crane	L.S.	1	166 844	166 844
2.2.6	Temporary survey jetty and platform	L.S.	1	91 795	91 795
			Bill total		453 887

Figure 28.   The direct cost of the project (3 of 7)

3   Earthworks   Bill No. 3

Item No.	Description	Unit	Quantity	Schedule rate	Amount
3.1	*Dredging*				
3.1.1	Dredging as specified on the drawings, including disposal of excavated soils	m³	3 000 000	0.435	1 305 000
3.1.2	Echo-sounding after the dredging work	L.S.	1	20 000.000	20 000
3.2	*Reclamation*				
3.2.1	Fill by hydraulic means	m³	70 000	0.306	21 420
3.2.2	Fill in the dry	m³	30 000	9.792	293 760
3.2.3	Levelling and topping by sand-gravel mixture	m²	52 500	1.275	66 937
3.3	*Slope protection*				
3.3.1	Supply, place and shape armour rocks 200–300 kg	m³	70 000	15.590	1 091 300
3.3.2	Supply, place in position filter fabric between rocks and the ground	m²	38 500	3.654	140 700
				Bill total	2 939 117

Figure 29. The direct cost of the project (4 of 7)

4   Unloading jetty Bill No. 4

Item No.	Description	Unit	Quantity	Schedule rate	Amount
4.1	*Piling*				
4.1.1	Supply of prestressed concrete piles Ø 750 mm, 125 mm thick, 20 m length	Nos.	96	960	92 160
4.1.2	Supply of prestressed concrete piles Ø 900 mm, 150 mm thick, 26.4 m length	Nos.	24	1 840	44 160
4.1.3	Supply of steel pipe piles Ø 300 mm, 9 mm thick, 18 m length with welding for joints	Nos.	42	550	23 100
4.1.4	Handle and drive vertical piles (Ø 750 × 125 t)	Nos.	96	3 374	323 904
4.1.5	Handle and drive batter piles (Ø 900 × 150 t)	Nos.	24	4 048	97 152
4.1.6	Handle and drive fender piles (Ø 300 × 9 t)	Nos.	42	1 690	70 980
4.1.7	Test piling Ø 750	Nos.	2	3 374	6 748
4.1.8	Test piling Ø 900	Nos.	2	4 048	8 096
4.1.9	Static loading test with maximum loading of 250 tonne, including temporary platform and reaction piles	Nos.	4	4 520	18 080
				Bill total	684 380

ESTIMATING THE COST OF CONSTRUCTION

Figure 29.   The direct cost of the project (4 of 7)

Figure 30.   The direct cost of the project (5 of 7)

4   Unloading jetty (*cont.*)   Bill No. 4

Item No.	Description	Unit	Quantity	Schedule rate	Amount
4.2	*Concrete work*				
	Chipping				
4.2.1	Chipping and disposal of pile top for Ø 750 piles	Nos.	96	134.630	12 925
4.2.2	Chipping and disposal of pile top for Ø 900 piles	Nos.	24	193.860	4 653
4.2.3	Cutting and disposal of pile top for Ø 300 piles	Nos.	42	150.000	6 300
	Pile caps				
4.2.4	Filling concrete with reinforcement cage for Ø 750 piles	Nos.	96	124.761	11 977
4.2.5	Filling concrete with reinforcement cages for Ø 900 piles	Nos.	24	133.079	3 194
	Deck slab concrete				
4.2.6	Deck slab concrete	m³	4 500	57.305	257 872
4.2.7	Curb (150 mm × 150 mm)	m³	7	61.305	429
	Reinforcement				
4.2.8	Reinforcement Grade 460 for 4.2.6 above:				
	(a) Up to 10 mm dia.	t	27	424.000	11 448
	(b) Up to 19 mm dia.	t	108	377.000	40 716
	(c) Up to 29 mm dia.	t	135	374.000	50 490
				Bill total	400 004

Figure 31.   The direct cost of the project (6 of 7)

4   Unloading jetty (*cont.*)   Bill No. 4

Item No.	Description	Unit	Quantity	Schedule rate	Amount
	Formwork				
4.2.9	Setting and removal of soffit formwork and falsework	m²	3 000	54.700	164 100
4.2.10	Setting and removal of side formwork	m²	450	54.700	24 615
4.2.11	Setting and removal of stop-end formwork	m²	180	54.700	9 846
4.3	*Fender system*				
4.3.1	Supply and installation of cell-type rubber fender	Nos.	42	3 240.000	136 080
4.3.2	Supply and installation of rubbing strip	Nos.	14	3 784.000	52 976
4.3.3	Anti-corrosion coating or painting upon the completion	L.S.	1	9 861.000	9 861
4.4	*Miscellaneous*				
4.4.1	Supply and installation of galvanized steel tube ladder	Nos.	2	1 412.500	2 825
4.4.2	Supply and installation of 50 t bollard	Nos.	12	385.930	4 631
4.4.3	Expansion joint: (a) galvanized steel angle 150 × 150 x 10	m	80	68.200	5 456
	(b) buffer plate	Nos.	4	185.000	740
				Bill total	411 130

Figure 32. The direct cost of the project (7 of 7)

5 Breakwater   Bill No. 5

Item No.	Description	Unit	Quantity	Schedule rate	Amount
5.1	*Caisson*				
5.1.1	Supply, place and shape rubble bedding stone (50–100 kg/pc)	m³	62 400	22.176	1 417 478
5.1.2	Supply, place and shape armour rocks (200–300 kg/pc)	m³	37 200	26.728	994 282
5.1.3	Supply, place in position filter fabric	m³	6 000	4.020	24 120
5.1.4	Inspection by sounding of mound formation	L.S.	1	10 000.000	10 000
5.2	*Concrete work*				
	In situ concrete				
5.2.1	Caisson concrete	m³	17 960	53.120	954 035
5.2.2	Topping concrete	m³	6 675	54.146	361 424
5.2.3	Non-shrinkage grout for gaps between caissons	Nos.	59	3 273.890	193 159
	Reinforcement				
5.2.4	Grade 460/425 for 5.2.1:				
	(a) up to 10 mm dia.	t	18	424.000	7 632
	(b) up to 19 mm dia.	t	1 257	377.000	473 889
	(c) up to 29 mm dia.	t	521	374.000	194 854
5.2.5	Grade 460/425 for 5.2.2:				
	(a) up to 10 mm dia.	t	33	424.000	13 992
	(b) up to 19 mm dia.	t	234	377.000	88 218
	(c) up to 29 mm dia.	t	67	374.000	25 058
				Bill total	4 758 141

Figure 33.   The direct cost summary sheet

The direct cost for the works =	
Bill No. 1	2 061 208
Bill No. 2	453 887
Bill No. 3	2 939 117
Bill No. 4	684 380
Bill No. 5	400 004
Bill No. 6	411 130
Bill No. 7	4 758 141
Total	11 707 867 dollars

3.4   The cost of site overheads

In addition to the direct cost of the works there will be the cost of the site overheads, those costs necessary to maintain the site whilst construction proceeds. These are categorized under the following headings:

(1)   Expenses for light and fuel.
(2)   Medical care and first aid.
(3)   Insurance.
(4)   Tax.
(5)   Salary.
(6)   Stationery.
(7)   Travel expenses.
(8)   Communications costs.
(9)   Accommodation.
(10)   Transportation.

This section provides a summary of these site overheads for the project together with examples of the detailed calculations behind these costs (see table 15).

(a)   *Examples of the site overhead calculations*

The following calculation examples show how the site overheads are calculated:

(1)   *Expenses for water and electricity*

Water

The cost of the desalination plant is covered within the mobilization costs. Additional monies need to be included to cover the storage and transportation of the water.

Cost of tanker lorry   =   600 dollars/month

Table 15. A summary of site overheads

Item	Local element	Offshore element	Total (dollars)
(1) Expenses for water and electricity	23 600	43 200	66 800
(2) Medical care and first aid	0	147 000	147 000
(3) Insurance			
– Car and third party insurance	0	180 480	180 480
– Workers' compensation	0	44 270	44 270
– Automobile insurance	3 634	0	3 634
(4) Tax			
– Corporation tax	100 000	0	100 000
– Income tax	74 933	0	74 933
(5) Salaries	23 958	504 000	527 958
(6) Stationery, etc.	16 900	0	16 900
(7) Travel expenses	29 429	48 000	77 429
(8) Communication	25 000	0	25 000
(9) Accommodation	89 200	28 000	117 200
(10) Transportation	13 000	76 000	89 000
Total	399 654	1 070 950	1 470 604

This is required for 18 months.

$$\text{Total cost} = 18 \times 600 \quad = \quad 10\ 800 \text{ dollars}$$
$$\text{Cost of storage tank} \quad = \quad 2\ 000 \text{ dollars}$$
$$\text{Total cost} = 10\ 800 + 2\ 000 \quad = \quad 12\ 800 \text{ dollars}$$

Electricity

Four generators are required at a hire rate of 600 dollars per month for the duration of the project.

$$\text{Total cost} = 600 \times 4 \times 18 \quad = \quad 43\ 200 \text{ dollars}$$
$$\text{Fuel cost} = 670\ 000 \text{ litres at } 0.18 \text{ dollars/litre}$$
$$= 10\ 800 \text{ dollars}$$
$$\text{Total electricity cost} = \quad 43\ 200 + 10\ 800$$
$$= \quad 66\ 800 \text{ dollars}$$

(2) *Medical and first-aid costs*

One doctor and one nurse will be required for the duration of the contract (18 months).

The cost of the doctor   = 4 000 dollars per month
The cost of the nurse   = 2 500 dollars per month
Equipment cost   = 30 000 dollars

Total medical and first-aid costs
= (4 000 × 18) + (2 500 × 18) + 30 000
= 72 000 + 45 000 + 30 000
= 147 000 dollars

(3) *Insurance*

Workers' compensation insurance is applicable on all local and UR workers and staff at a premium of 3.3 per cent of their total wage and salary costs. This gives a total premium of 3.3 per cent × 1,341,500 = 44,270 dollars.

Automobile insurance quotations are obtained for the cars and minibuses to be used on the project. These quotations are obtained from local insurance companies by the local agent. The total cost = 4,485 dollars per annum. Total cost of the project = 6,728 dollars (assumes 18 months).

Contractor's all risks insurance and third party insurance for the project is obtained at a total cost of 180,480 dollars.

Total insurance costs =   180 480 + 6 728 + 44 270
                       =   231 478 dollars

(4) *Taxes*

Reference should be made to clause 73 of the conditions of contract. The contractor is liable for the following taxes:

– an income tax element payable on the staff and worker salaries; and
– corporation tax on the profit made by the company on the project.

The income tax element is estimated by calculating the average salary cost and average wage cost on the project, and then calculating the likely income tax based upon the current Federalstani laws. This gives a figure of 74,933 dollars.

The company's anticipated tax liability under Federalstani law is estimated to be a maximum of 100,000 dollars based upon the anticipated profit on the project and the current tax situation with respect to non-resident companies employed in the construction of capital projects.

Therefore the total taxes to be paid =   74 933 + 100 000
                                      =   174 933 dollars.

(5) *Salaries*

Fourteen staff, plus locally-recruited staff, required for the 18-month duration of the project.

	Unit rate	Unit	Quantity	Amount
Staff	2 000	Nos.	252	504 000
Local staff	150	Nos.	54	8 100
Supervisor	125	Nos.	90	11 250
Office boy	32	Nos.	36	1 152
Watchman	32	Nos.	108	3 456
				527 958

(6) *Stationery*

	Unit rate	Unit	Quantity	Amount
Copy machine	2 000	Nos.	2	4 000
Camera, etc.	400	Nos.	4	1 600
Desk	100	Nos.	30	3 000
Table	80	Nos.	10	800
Bookshelf	100	Nos.	20	2 000
Locker	150	Nos.	20	3 000
Drawing equipment		Nos.	1	500
Papers, etc.		Nos.	1	2 000
				16 900

(7) *Travel expenses*

These expenses will apply to the staff, medical staff and maximum of 32 workers.

	Unit	Quantity	Amount
Residential work permits	Item	1	14 429
Flight charge (return) 1 000	Nos.	1	48 000
Other travel	L.S.	1	10 000
Domestic travel	L.S.	1	5 000
			77 429

(8) *Communication*

	Unit	Quantity	Amount
Telephone	L.S.	1	15 000
Telex	L.S.	1	5 000
Broadcast on site	L.S.	1	2 000
Walkie-talkie	L.S.	1	3 000
			25 000

(9) *Accommodation*

	Unit rate	Unit	Quantity	Amount
Houses	50	m²	1 000	50 000
Wiring		L.S.	1	2 000
Plumbing		L.S.	1	10 000
Air-conditioning		L.S.	1	20 000
Appliances		L.S.	1	10 000
				92 000

Canteen (site and residence)

	Unit rate	Unit	Quantity	Amount
Shed	50	m²	100	5 000
Refrigerator	2 000	Nos.	4	8 000
Table, chair		Nos.	1	3 000
Miscellaneous		Nos.	1	1 000
Catering		L.S.		8 000
				25 000
		Total		117 000

(10) *Transportation*

	Unit rate	Unit	Quantity	Amount
Pick-up truck	7 000	Nos.	4	28 000
Minibus	12 000	Nos.	4	48 000
Driver	50	Nos.	60	3 000
Petrol		L.S.	1	10 000
				89 000

3.5 The total construction cost

The total construction cost is the sum of the direct cost and the site overheads calculated in the previous section.

Construction cost   = direct cost + site overheads
     = 11 707 867 + 1 470 604
     = 13 178 471 dollars

Table 16 shows the breakdown of this sum into "local" and "offshore" currency elements.

Table 16. Total construction costs – local and offshore currency elements

Description	Federalstan (local) element	UR (offshore) element	Total (dollars)
Direct cost	5 638 262	6 069 605	11 707 867
Site overheads	399 654	1 070 950	1 470 604
	6 037 916 (45.8%)	7 140 555 (54.2%)	13 178 471 (100%)

THE DECISION ON THE TENDER PRICE AND THE SUBMISSION OF THE TENDER

4

4.1 The decision on the tender price

At the tender meeting a few days before the tender is due to be submitted, the Managing Director of the company together with the Chief Estimator and the staff who have been directly involved with the preparation of the estimate meet to discuss the estimate that has been prepared and decide upon the final tender to be submitted.

Prior to the meeting the estimators have prepared the final calculated cost for the construction of the works together with a detailed breakdown of the project costs. It is the responsibility of the meeting to decide the following:

- any adjustments to be made to the construction cost estimate;
- the allowance to be made for head office overheads;
- the allowance to be made to cover the risk element in the project;
- the required profit; and
- the total tender sum and allocation of indirect costs.

The Managing Director confirms that the company will be actively pursuing this contract. The company has been seeking work in the region for some time and this project represents a good opportunity for the company to establish itself in this geographical region.

(a) Adjustments to the construction cost estimate

The Managing Director then asks the estimator to review the project estimate in detail and in particular to state any assumptions made in the preparation of the estimate. The planning engineer is asked to explain fully the method statement and the project programme.

It is decided that the following adjustments will be made to the estimate:

  (i) A sum of 664,800 dollars will be *added* to the estimate, being the hire cost of the plant to be transported from the UR during the mobilization period. (See section 3.3.)

(ii)   A sum of 60,000 dollars will be *deducted* from the estimate as it is believed that the piling works will be completed at a faster rate than anticipated in the method statement. This will release some hired plant from the project earlier than expected.

(iii)   The supply of the armour rock and sand for the project has been subcontracted to a local supplier. The estimator has indicated that an alternative supplier has just contacted the company with an offer to supply the rock and sand at a lower rate. It is decided to include this rate within the estimate. Consequently a total of 67,500 dollars will be *deducted* from the construction cost.

(iv)   The prices included in the estimate for reinforcement were those available at the start of the estimate. The company's agent in Federalstan has been in contact in the last few days to inform them that these prices are expected to rise in the near future. It is decided to *add* a sum of 59,500 dollars to cover this price increase.

(v)   A review of the estimate shows that miscellaneous savings of 25,560 dollars may be expected. This sum is therefore *deducted* from the estimate.

(b)   The allowance for head office overheads

The Managing Director indicates that the project must be able to contribute 750,000 dollars towards expected head office costs over the 18-month period of the project. It is decided that the location of the project will require additional head office staff to be employed for the duration of the project. These staff will liaise with the site and ensure that sufficient back-up is provided to ensure that construction operations proceed smoothly. An additional sum of 75,000 dollars is included to cover the costs of these staff.

The project will require the provision of both a tender bond and a performance bond. The total cost of these bonds must be met by the project. The cost of securing these bonds is estimated at some 70,000 dollars by the company's finance director. This sum is included as a head office overhead.

Included within the head office costs must be an allowance for any interest charges to be paid by the company in securing working capital to complete the project. This normally requires a full cash flow to be produced for the project. However, for this project, the cash flow is not necessary because the client's agreement to an initial prepayment results in sufficient monies being available to fund the construction work.

Therefore the total head office overheads to be included in the estimate are:

$$750\ 000 + 75\ 000 + 70\ 000 = 895\ 000 \text{ dollars.}$$

(c) The allowance to cover risk

An assessment is made of the element of risk involved in the project so that an allowance may be made in the tender to cover this risk. This includes a consideration of the physical and technical uncertainties relating to the project together with the commercial risks involved.

The following factors are included in the discussions by the tender adjudication panel:

- the contract period;
- payment terms and taxation requirements;
- bonding arrangements;
- import tax drawback and refunds;
- the specification for the works;
- the physical aspects of the site;
- the supply of local labour, plant and materials;
- currency fluctuations;
- inflation; and
- penalties for non-performance and delay.

It is decided to add a total of 775,000 dollars to the tender to cover all these risk elements.

(d) The required profit margin

The Managing Director indicates that a profit of 3 per cent on this project would be acceptable to the Board. This has been previously decided by the Board following consideration of: the desire to win the contract; the current workload; and anticipated orders for other work.

(e) The total tender sum and the allocation of indirect costs

The total tender sum for the project is therefore as shown below:

Direct cost	11 707 867
Site overheads	1 470 604

Tender adjudication adjustments

Hire plant addition	664 800
Piling adjustment	(60 000)
Rock supply	(67 500)
Reinforcement costs	59 500
Miscellaneous savings	(25 560)
Amended construction cost	13 749 711

plus

Head office overheads	895 000
Allowance for risk	775 000
Profit allowance	412 491
(3% of amended	
construction cost)	
Total tender sum	15 832 202

Sum to be added to the direct cost total
= 15 832 202 – 11 707 867
= 4 124 335

It is decided by the Managing Director and the tender adjudication panel to apportion this sum throughout the bill by:

– adding 25 per cent to each bill item;
– adding the remaining sum to Bill item 1.1, Mobilization and demobilization; and
– converting the dollar rates to Federalstani Nu.

4.2 The submission of the tender

As stated in contract documents shown in Chapter 4, the contractor is required to deliver the tender by a stated time and date to a specified address. Any tender received after that deadline will be rejected and returned without opening.

The tender submission must comprise:
(i)   Form of Tender
(ii) Bill of Quantities
(iii) A Bid Bond completed by the contractor's guarantor.

Figure 34 shows the completed Form of Tender.

Figures 35 to 41 show the Bill of Quantities priced with the tender rates in Federalstani Nu.

Figure 42 shows the Tender Summary Sheet.

Figure 43 shows the completed Bid Bond.

Note: In the Tender Summary Sheet the contractor is required to enter the amount within each section of the bill that is required to be paid in foreign currency. The sums entered on the Tender Summary Sheet represent the contractor's decision after considering all the risk elements involved relating to the payment in Federalstani Nu and UR dollars. The minimum amount to be paid in Federalstani Nu is the construction costs to be paid in local currency.

Figure 34.   The completed Form of Tender Document

Form of Tender

Federalstan Harbour Authority,
Contract Road, Capital City,
FEDERALSTAN.

GENTLEMEN

Having examined the Tender Documents for the above Contract, including Drawings, Conditions of Contract, Specifications and Bill of Quantities for the construction of the above-mentioned Works, the receipt of which is hereby acknowledged, we offer to construct and maintain the whole of the Works as described in, and in accordance with, the said Tender Documents including Addenda Nos._____N/A_____, for the sum of _____474,966,060_____Nu.

We undertake, if our Tender is accepted, to commence the Works within _____30_____days and to complete and deliver the Works in accordance with the Contract within _____540_____days calculated from the date of commencement of the Works and in accordance with the Time Schedule.

If our Tender is accepted we will provide the Performance Bond in the sum of _____47,496,606 Nu_____, equal to 10 per cent of the Contract Price, for the due performance of the Contract.

We agree to abide by this Tender for the period of one hundred and twenty (120) days from the date fixed for Tender Closing pursuant to clause 14 of the Instructions to Tenderers and it shall remain binding on us and may be accepted at any time before the expiration of that period.

We attach the Appendices to the Tender Form, duly completed and signed.

Until a formal Contract is prepared and executed, this Tender together with your written acceptance thereof by your notification of Award shall constitute a binding contract between us.

We understand that you are not bound to accept the lowest or any Tender that you may receive.

We are, Gentlemen,

Yours faithfully,

Signature: _____

Address: UR Construction

Victoria Street, Kingston

Date: 1 March 1994

Note: This Form of Tender is adapted from the FIDIC Conditions of Contract.

Appendix to the Form of Tender

Details in this Appendix are cross-linked to the Conditions of Contract by the clause numbers shown.

	Clause	
Amount of Bond	10	10 per cent
Minimum amount of third party insurance	23	60 million Nu
Period of commencement	41	Within thirty (30) days after receipt of Engineer's order to commence
Time for completion	43	540 days calculated from the last day of the period of commencement
Limit of liquidated damages	47	10 per cent of Contract Price
Period of maintenance	47	365 days
Percentage of retention	60	10 per cent
Minimum amount of interim certificate	60	300 000 Nu
Time within which payments to be made after issue of certificates	60	30 days

Note: This Form of Tender is adapted from the FIDIC Conditions of Contract.

Figure 35.   The tender rates for the project (1 of 7)

1   General item   Bill No. 1

Item No.	Description	Unit	Quantity	Schedule rate (Nu)	Amount (Nu)
1.1	Mobilization of equipment to the site and demobilization	L.S.	1	101 569 400.000	101 569 400
1.2	*Topograpic survey*				
1.2.1	Survey on land	L.S.	1	375 000.000	375 000
1.2.2	Sounding offshore	L.S.	1	937 500.000	937 500
1.3	*Soil investigation*				
1.3.1	Onshore boring, including samples and field test	m	50	5 625.000	281 250
1.3.2	Offshore boring, including temporary staging	m	100	5 625.000	562 500
1.3.3	Laboratory test on the soil samples, including preparation of reports	L.S.	1	375 000.000	375 000
1.4	*Provide and maintain the following for exclusive use of the engineers*				
1.4.1	Offices as specified in clause 102(4)	L.S.	1	1 147 500.000	1 147 500
1.4.2	New vehicles for their transportation	Nos.	5	315 000.000	1 575 000
1.4.3	New generators (300KVA)	Nos.	2	1 575 000.000	3 150 000
1.4.4	Potable water, oil and their storage tanks	L.S.	1	900 000.000	900 000
1.4.5	Detached houses for accommodation	Nos.	5	468 750.000	2 343 750
				BILL TOTAL	113 216 900

Figure 36. The tender rates for the project (2 of 7)

2 Site preparation   Bill No. 2

Item No.	Description	Unit	Quantity	Schedule rate (Nu)	Amount (Nu)
2.1	*Site clearance*				
2.1.1	General site clearance	L.S.	1	505 620.000	505 620
2.1.2	Disposal area	L.S.	1	48 630.000	48 630
2.2	*Preparation of site*				
2.2.1	Site office with gate and fencing	L.S.	1	1 500 000.000	1 500 000
2.2.2	Site laboratory for tests of soil, cement, water, etc., with equipment and staffing	L.S.	1	1 635 000.000	1 635 000
2.2.3	Warehouses for all materials used for the permanent works	L.S.	1	919 050.000	919 050
2.2.4	Desalination plant for potable water for living and construction purposes	L.S.	1	2 713 500.000	2 713 500
2.2.5	Temporary loading jetty and gantry crane	L.S.	1	6 256 650.000	6 256 650
2.2.6	Temporary survey jetty and platform	L.S.	1	3 442 320.000	3 442 320
				BILL TOTAL	17 020 470

INTERNATIONAL BIDDING CASE STUDY

Figure 37. The tender rates for the project (3 of 7)

3 Earthworks   Bill No. 3

Item No.	Description	Unit	Quantity	Schedule rate (Nu)	Amount (Nu)
3.1	*Dredging*				
3.1.1	Dredging as specified on the drawings, including disposal of excavated soils	m^3	3 000 000	16.3125	48 937 500
3.1.2	Echo-sounding after the dredging work	L.S.	1	750 000.000	750 000
3.2	*Reclamation*				
3.2.1	Fill by hydraulic means	m^3	70 000	11.475	803 250
3.2.2	Fill in the dry	m^3	30 000	367.200	11 016 000
3.2.3	Levelling and topping by sand-gravel mixture	m^2	52 500	47.812	2 510 130
3.3	*Slope protection*				
3.3.1	Supply, place and shape armour rocks 200–300 kg	m^3	70 000	584.625	40 923 750
3.3.2	Supply, place in position filter fabric between rocks and the ground	m^2	35 000	150.750	5 276 250
				BILL TOTAL	110 216 880

Figure 38. The tender rates for the project (4 of 7)

4 Unloading jetty   Bill No. 4

Item No.	Description	Unit	Quantity	Schedule rate (Nu)	Amount (Nu)
4.1	*Piling*				
4.1.1	Supply of prestressed concrete piles ⌀ 750, 125 mm thick 20 m length	Nos.	96	36 000.000	3 456 000
4.1.2	Supply of prestressed concrete piles ⌀ 900, 150 mm thick 26.4 m length	Nos.	24	69 000.000	1 656 000
4.1.3	Supply of steel pipe piles ⌀ 300, 9 mm thick, 18 m length with welding for joints	Nos.	42	20 625.000	866 250
4.1.4	Handle and drive vertical piles (⌀ 750 × 125 t)	Nos.	96	126 525.000	866 250
4.1.5	Handle and drive batter piles (⌀ 900 × 150 t)	Nos.	24	151 800.000	3 643 200
4.1.6	Handle and drive fender piles (⌀ 300 ×9 t)	Nos.	42	63 375.000	2 661 750
4.1.7	Test piling ⌀ 750	Nos.	2	126 525.000	253 050
4.1.8	Test piling ⌀ 900	Nos.	2	151 800.000	303 600
4.1.9	Static loading test with max. loading of 250 tonne, including temporary platform and reaction piles	Nos.	4	169 500.000	678 000
				BILL TOTAL	25 664 250

Figure 39.   The tender rates for the project (5 of 7)

4   Unloading jetty (*cont.*)   Bill No. 5

Item No.	Description	Unit	Quantity	Schedule rate (Nu)	Amount (Nu)
4.2	*Concrete work*				
	Chipping				
4.2.1	Chipping and disposal of pile top for ∅ 750 piles	Nos.	96	5 049.062	484 710
4.2.2	Chipping and disposal of pile top for ∅ 900 piles	Nos.	24	7 270.000	174 480
4.2.3	Cutting and disposal of pile top for ∅ 300 piles	Nos.	42	5 625.000	236 250
	Pile caps				
4.2.4	Filling concrete with reinforcement cage for ∅ 750 piles	Nos.	96	4 678.438	449 130
4.2.5	Filling concrete with reinforcement cages for ∅ 900 piles	Nos.	24	4 991.250	119 790
	Deck slab concrete				
4.2.6	Deck slab concrete	m^3	4 500	2 148.993	9 670 200
4.2.7	Curb (150 mm × 150 mm)	m^3	7	2 297.143	16 080
	Reinforcement				
4.2.8	Reinforcement grade 460 for 4.2.6 above:				
	(a) up to 10 mm dia.	t	27	15 900.000	429 300
	(b) up to 19 mm dia.	t	108	14 137.500	1 526 850
	(c) up to 29 mm dia.	t	135	14 024.889	1 893 360
				BILL TOTAL	15 000 150

Figure 40.   The tender rates for the project (6 of 7)

4   Unloading jetty (*cont.*)   Bill No. 4

Item No.	Description	Unit	Quantity	Schedule rate (Nu)	Amount (Nu)
	Formwork				
4.2.9	Setting and removal of soffit formwork and falsework	m²	3 000	2 051.250	6 153 750
4.2.10	Setting and removal of side formwork	m²	450	2 051.270	923 070
4.2.11	Setting and removal of stop-end formwork	m²	180	2 051.330	369 240
4.3	*Fender system*				
4.3.1	Supply and installation of cell-type rubber fender	Nos.	42	121 500.000	5 103 000
4.3.2	Supply and installation of rubbing strip	Nos.	14	141 900.000	1 986 600
4.3.3	Anti-corrosion coating or painting upon the completion	L.S.	1	369 780.000	369 780
4.4	*Miscellaneous*				
4.4.1	Supply and installation of galvanized steel tube ladder	Nos.	2	52 965.000	105 930
4.4.2	Supply and installation of 50 t bollard	Nos.	12	14 472.500	173 670
4.4.3	Expansion joint:				
	(a) galvanized steel angle 150 × 150 × 10	m	80	2 557.500	204 600
	(b) buffer plate	Nos.	4	6 937.500	27 750
				BILL TOTAL	15 417 390

Figure 41. The tender rates for the project (7 of 7)

5   Breakwater   Bill No. 5

Item No.	Description	Unit	Quantity	Schedule rate (Nu)	Amount (Nu)
5.1	*Caisson mound*				
5.1.1	Supply, place and shape rubble bedding stone (50–100 kg/pc)	m³	62 400	851.850	53 155 440
5.1.2	Supply, place and shape armour rocks (200–300 kg/pc)	m³	37 200	1 002.301	37 285 590
5.1.3	Supply and place in position filter fabric	m²	6 000	150.750	904 500
5.1.4	Inspection by sounding of mound formation	L.S.	1	375 000.000	375 000
5.2	*Concrete work*				
	In situ concrete				
5.2.1	Caisson concrete	m³	17 960	1 992.000	35 776 320
5.2.2	Topping concrete	m³	6 675	2 030.472	13 553 400
5.2.3	Non-shrinkage grout for gaps between caissons	Nos.	59	122 770.678	7 243 470
	Reinforcement				
5.2.4	Grade 460/425 for 5.2.1:				
	(a) up to 10 mm dia.	t	18	15 900.000	286 200
	(b) up to 19 mm dia.	t	1 257	14 137.490	17 700 830
	(c) up to 29 mm dia.	t	521	14 024.971	7 307 010
5.2.5	Grade 460/425 for 5.2.2:				
	(a) up to 10 mm dia.	t	33	15 900.000	524 700
	(b) up to 19 mm dia.	t	234	14 137.436	3 308 160
	(c) up to 29 mm dia.	t	67	14 024.776	939 660
				BILL TOTAL	178 430 020

Figure 42. The Tender Summary Sheet

	GRAND SUMMARY		Total (Nu)
Bill No. 1	Sheet 1		
	(Foreign currency element	86.48%)	113 216 900
Bill No. 2	Sheet 2		
	(Foreign currency element	37.71%)	17 020 470
Bill No. 3	Sheet 3		
	(Foreign currency element	57.84%)	110 216 880
Bill No. 4	Sheet 4		
	(Foreign currency element	84.78%)	25 664 250
	Sheet 5		
	(Foreign currency element	31.74%)	15 000 150
	Sheet 6		
	(Foreign currency element	37.73%)	15 417 390
Bill No. 5	Sheet 7		
	(Foreign currency element	52.20%)	178 430 020

Note: All foreign currency payments shall be in UR dollars.

Signed: _____

On behalf of

Tenderer: _____

Address: _____

_____

Total tender sum   474 966 060

Figure 43. The completed Bid Bond

Form of Bid Bond

Whereas _____ UR Construction _____ (hereinafter called Tenderer) has submitted its Tender dated _____ 1 April 1992 _____ for the construction of a new cargo jetty and breakwater (hereinafter called the Tender) _____ know all men by these present that WE the Bank of ____ UR ____ of ____ Kingston, UR ____ having our registered office at ____ Main Street, Kingston ____ (hereinafter called the Bank) are bound to Federalstan Harbour Authority (hereinafter called the Employer) in the sum of ____ 23,748,303 ____ Nu for which payment to be made to the said Employer, the Bank binds itself, its successors and assigns by these present. Sealed with the Common Seal of the said Bank this ____ 30th ____ day of ____ March ____ 1992.

The CONDITIONS of this obligation are:

1. If the Tenderer withdraws its Tender during the period of Tender validity specified by the Tenderer on the Tender form; or

2. If the Tenderer, having been notified of the acceptance of its Tender by the Employer during the period of its Tender validity:

   (a) fails or refuses to execute the Contract Form when requested; or

   (b) fails or refuses to furnish the Performance Bond, in accordance with the Instructions to Tenderers.

We undertake to pay to the Employer up to the above amount according to and upon receipt of its first written demand, without the Employer having to substantiate its demand, provided that in its demand the Employer shall note that the amount claimed by it is due to it owing to the occurrence of one or both of the two above-stated conditions, specifying the occurred condition or conditions.

This guarantee will remain in force up to and including thirty (30) days after the period of Tender validity, and any demand in respect thereof should reach the Bank not later than such date.

Name of Bank: the Bank of UR

Signature of authorized
Representative of Bank: _____

Signature of Witness: _____

Name of Witness: _____

Address: _____

                                        _____

Note: This Form of Bid Bond is adapted from the FIDIC Conditions of Contract.